Water and Disasters: Cases from the High Level Experts and Leaders Panel on Water and Disasters

In Focus – Special Book Series

Water and Disasters: Cases from the High Level Experts and Leaders Panel on Water and Disasters

Editors
Jerome Delli Priscoli and Kenzo Hiroki

IWA PUBLISHING

Published by **IWA Publishing**
Alliance House
12 Caxton Street
London SW1H 0QS, UK
Telephone: +44 (0)20 7654 5500
Fax: +44 (0)20 7654 5555
Email: publications@iwap.co.uk
Web: www.iwapublishing.com

First published 2019

British Library Cataloguing in Publication Data
A CIP catalogue record for this book is available from the British Library

ISBN: 9781789060782 (paperback)
ISBN: 9781789060799 (eBook)
ISBN: 9781789060805 (ePUB)

Contents

Editorial

Water and Disasters: Cases from the High Level Experts and Leaders Panel on Water and Disasters

Jerome Delli Priscoli[a] and Kenzo Hiroki[b]

[a]*TEC Chair Global Water Partnership (GWP), Senior Advisor USACE IWR (ret.), Editor in Chief Water Policy*
[b]*Professor, National Graduate Institute for Policy Studies (GRIPS), Coordinator, High Level Experts and Leaders Panel on Water and Disasters (HELP)*

Water is life. But water is also a threat to life. During the past decade, the risks from water-related disasters are increasing and hamper sustainable development by causing political, social, and economic upheaval in many countries. Water-related disasters, such as floods, droughts, storm surges and tsunamis, account for 90% of all disasters in terms of number of people affected.

The issue of "water and disasters" must be addressed if we hope to make sustainable development a reality. We must share our experiences and lessons learned, strengthen regional coordination and collaboration, and set common goals and targets in order to lay a foundation for weathering the water-related disasters to come, and make progress towards creating a better-prepared and more resilient society. All these elements should be translated into clear-cut messages and practical advice for decision makers to create effective policies and mechanisms that address water and disaster issues appropriately at all levels.

The High Level Experts and Leaders Panel on Water and Disasters (HELP) was established to assist the international community, governments and stakeholders in mobilizing political will and resources. It will promote actions to raise awareness, ensure coordination and collaboration, establish common goals and targets, monitor progress, and take effective measures aimed at addressing the issues of water and disasters.

During 2014, members of HELP solicited case studies on water and disasters. After review, these contributions comprise this special issue of *Water Policy*. The cases and the special issue were also done to provide input to discussion on this topic at the World Water Forum 7 in the Republic of South Korea.

The special issue consists of 10 cases studies and concludes with a summary position paper of members of the HELP: 'Water-Related Disaster Risk Reduction: Time for Preventive Action: Position paper of the High Level Experts and Leaders Panel on Water and Disasters'.

The special Issue begins with 'A Global Perspective on Flood Disasters' written by Dr Wolfgang Kron of the Munich Reinsurance Company (Kron, 2015). The article notes that various disasters in recent decades have confirmed that the risk from water-related events has been increasing significantly worldwide. The driving factors of this risk are the unabated increase in global population, the

doi: 10.2166/wp.2015.000

concentration of people in high-risk areas such as coasts, flood plains and hillsides, the rise in vulnerability of assets, infrastructure and social systems, and the consequences of environmental and climatic changes. The article concludes that risk reduction requires general awareness at all levels of society and a partnership between the public sector, the people concerned and the insurance industry. It also notes that structural and non-structural precautionary measures are always cheaper in the long run than paying losses and that overall economic consequences are significantly less severe in societies with a high insurance penetration.

Specific cases follow this global overview and begin with a Japanese analysis of 'Lessons Learned from Mega-Disasters and Future Policy Development on Water-related Disaster Management in Japan' (Tachi, 2015). The article presents lessons learned from the Great East Japan Earthquake and Tsunami, a disaster caused by a low-probability but extremely large hazard, and introduces the distinct features of Japanese water-related disaster management, including those enhanced and strengthened based on the recent lessons. Finally, drawing from Japan's experiences, it offers messages to be sent to the world from the water-related disaster community.

An article on the world renowned Dutch Delta works, 'The Deltaplan revisited: Changing perspectives in the Netherlands' flood risk reduction philosophy', follows (Schultz van Haegen & Wieriks, 2015). It was prepared by Minister Melanie Schultz van Haegen, and Koos Wieriks. The article notes that worldwide, the risk of weather-related disasters, such as floods and landslides, has increased strongly over the last decades. This has caused the Netherlands to adjust their policies to water disaster reduction. The example of the changing perspectives in the Netherlands' flood risk reduction philosophy illustrates the concepts of integrated disaster risk reduction to water-related disasters in general. It notes that measures can be designed based on future scenarios and a flexible, adaptive approach that allows for switching between strategies along adaptation pathways when needed in view of socioeconomic developments or climate change is necessary. It concludes that disaster risk reduction; Water Resources Management and Climate Adaptation should no longer be treated as separate topics but should be merged into an integrated approach. It calls for a further elaboration of the Hyogo framework, the Sustainable Development Goals (including SDGs on water and urbanization), the UNFCCC, and Habitat III. It finally cautions to make sure that the boundary conditions for the implementation of the necessary measures to reduce risk and increase resilience are in place, such as long-term funding, the elements of fair governance, and stakeholder participation.

Drs Jerome Delli Priscoli and Eugene Stakhiv of the US Army Corps of Engineers follow with 'Water-Related Disaster Risk Reduction (DRR) Management in the United States: Floods and Storm Surges' (Delli Priscoli & Stakhiv, 2015). The article combines a review of history, concepts and comparison of lesson from three recent cases of Hurricane Katrina, 2011 Mississippi floods and Hurricane Sandy. Reviewing history, this article describes a US federal system that presents major challenges to coordinating water resources development and DRR, at both the watershed level and metropolitan area scales.

This article reviews the performance of existing flood protection systems of three recent disasters; Hurricane Katrina (2005), Superstorm Sandy (2012) and the Mississippi River flood (2011), and the effectiveness of post-flood reconstruction plans. These three events provide insights into the evolution of the federal system and how the 'federal interest' was and is being implemented, and what it portends for future management. The cases revealed new vulnerabilities and weaknesses in the US DRR responses and planning, while contrasting the relative successes of long term, strategic DRR planning and investments in the case of the Mississippi River and Tributaries system. The article analyses this history and recent cases primarily from the perspective of the US Army Corps of Engineers, which has been a major federal actor and proponent of these policies throughout US history, with varying degrees of implementation authority in each of the three cases.

Floods and typhoons are two of the greatest water disasters affecting South East Asia. Keizrul Abdullah *et al.*, in a 'Tale of Three Cities', notes that the rapid pace of development in the South East Asia region has resulted in a disproportionate increase of the runoff and a many fold increase in river discharges leading to more frequent and more intense flooding (Abdullah *et al.*, 2015). This situation is expected to be further aggravated due to the impact of global warming and climate change. To cope with such challenges, countries in South East Asia are developing their policy responses tailored to suit their local conditions and environment.

This article looks at the water disaster situation and the policy responses in three cities in South East Asia. The cities are Bangkok, Kuala Lumpur and Metro Manila, the capital cities of the Kingdom of Thailand, the Federation of Malaysia and the Republic of the Philippines, respectively. Although all three countries are in the same climatic zone, i.e. the monsoonal humid tropics, due to their geographical locations, water disasters impact differently on them and so too the remedial measures differ.

Wolfgang Grabs and Hans Moser of the Federal Institute of Hydrology, Koblenz, Germany and the International Commission for the Hydrology of the Rhine Basin, Lelystad, The Netherlands, contributed an article entitled 'Elbe Floods Translating Policies into Actions – The Case of the Elbe River' (Grabs & Moser, 2015). The article describes methods and processes to link policy development to the implementation of these policies in actionable implementation plans. It is shown that policies can only be implemented effectively if they are embedded in a legal framework that is designed to facilitate achievement of the policy objectives.

Similar to the situation described in the US article, the article shows different levels of policy-making and decision-support for the development of policies at different levels, ranging from the level of Federal States in Germany to the trans-boundary level and policy development and implementation at the level of the European level in the framework of the European Framework Directive. Using the Elbe River as a case study, the paper shows the need for anchoring regional, trans-boundary and state level policies to mandated national institutions.

A key lesson learnt from the Elbe river basin is that policy integration is of utmost importance: policies need to be developed and existing policies be adapted at different levels across sectors in such a way that these policies are mutually complementary, non-contradictory and implementable to achieve improved integrated flood risk management. Action Plans that foster the implementation of policies need to be flexible to respond to changing conditions and the availability of resources for their implementation. The paper also demonstrates that a balance needs to be reached with regard to structural and non-structural measures in flood risk management to arrive at a truly integrated flood risk management strategy and its implementation. The development of research policies on the basis of sound science is indispensable in support of policy development and its implementation.

Martine Grambow *et al.*, in 'Lessons Learned from the Big Floods, 1993–2013 – the Bavarian Flood action program 2020 plus', note that flood protection has historically been critical to society's advancement (Grambow *et al.*, 2015). Today they state that against the background of meeting the partially geopolitical and global changes from growing vulnerability, the system of natural hazard management must be audited on a regular basis. This is illustrated using the example of Bavaria. The lessons from the big floods in the period of 1999–2013 led to a fundamental modification in the existing integral flood protection strategy of Bavaria. Concepts of flood risk management and resilience, dealing with extreme flood events, which exceed the design discharge of protection systems, came to the fore. That includes management of flash floods or floods, handling of potential retention areas, inclusion of insurance or resettlement or the burden of maintenance are further challenges, which are briefly addressed in the following article.

The Bavarian case is followed by 'Losing memory – the risk of a major flood in Paris region – Improving prevention policies' (Baubion, 2015). This article provides a snapshot of the key findings of the OECD Review on flood risk prevention policies in the Paris metropolitan area. With an innovative flood risk analysis model, the study shows that a major flooding of the Seine River similar to the flood disaster of 1910 could affect up to 5 million residents in the greater Paris area and cause up to 30 billion Euros worth of damage. Economic growth, jobs and public finances could also be significantly affected at the national level. The OECD Review on Flood Risk Management of the Seine River – commissioned by the Basin Organization Seine Grand Lacs with the French Ministry of Ecology and Île-de-France regional council – recommends raising risk awareness among citizens and businesses and improving the resilience of the metropolitan area to flood risks. Recent floods in Europe and New York City's Hurricane Sandy disaster in 2012 illustrated the vulnerability of today's ever-denser cities to flooding and the need to adapt critical infrastructure systems to be able to cope with extreme weather events. The 1910 Paris flood took several weeks to subside. The absence of a major flood in the Paris region for the past half century means the spectre has largely faded from collective memory. Yet population growth and the density of transport and energy infrastructure that has grown up around the French capital mean the area is greatly exposed to flood risks, despite the upstream dam-reservoirs and other defences now in place. The OECD review suggests ways to minimise the risks and better prepare the Île-de-France region. It notes that proposed projects to develop and expand the city's transport and logistics networks offer an opportunity to put some of its suggestions into practice.

In the article 'Integrated policies and practices for flood and drought risk management', Grobicki *et al.* (2015) offer a perspective form the Global Water Partnership (GWP). The article argues for an integrated approach to the management of water-related disasters that becomes a full part of the political decision-making process at the earliest possible moment and focuses on preparedness, mitigating the negative impacts and also considering the positive impacts, particularly of floods. By doing this there is an opportunity to consider the three pillars of sustainable development, and understand the options that exist and the trade-offs that may need to be made between economic efficiency, environmental sustainability and social equity. Within the post-2015 agenda, water-related disasters are addressed by targets under a number of different Sustainable Development Goals. It says that there is an urgent need to build disaster-resilient societies through more integrated policies and practices, including stakeholders' perspectives and a partnership approach. The article provides comparative stakeholder perspectives and approaches from around the world that are putting these ideas into practice.

In the final article Bruna Mendonca *et al.*, in 'Integrated Actions in the Management of Critical Hydrologic Events: Flood Vulnerability Atlas and Situation Rooms', describe an operational approach now taken in Brazil (De Sá E Mendonça & De Abreu, 2015). The authors note that due to the increase in urban populations, there was also an increased incidence of critical events, especially those related to floods and landslides. Since the implementation of the National Civil Defense System in Brazil, the focus on disaster management is no longer just to those affected. It has become one of managing disaster risk, in addition to the response, including prevention and minimizing the effects of the critical event. The article describes the expansion of a Hydrological Monitoring Network, in particular the Alert Network, the completion of the Atlas of Vulnerability to Floods and deployment of Situation Rooms in the states. These are all meant to strengthen the role of disaster risk management and enable greater responsiveness to these events and use the information gathered as a water resources management tool.

The Special Issue concludes with a short statement of the members of the HELP entitled 'Water-related Disaster Risk Reduction: time for preventive action' (Wieriks & Vlaanderen, 2015). It is a position paper of the High Level Experts and Leaders Panel on Water and Disasters.

References

Abdullah, K., Anukularmphai, A., Kawasaki, T. & Nepomuceno, D. (2015). A tale of three cities: water disaster policy responses in Bangkok, Kuala Lumpur and Metro Manila. *Water Policy 17*(suppl.1), 89–113.

Baubion, C. (2015). Losing memory – the risk of a major flood in the Paris region: Improving prevention policies. *Water Policy 17*(suppl.1), 156–179.

Delli Priscoli, J. & Stakhiv, E. (2015). Water-related disaster risk reduction (DRR) management in the United States: floods and storm surges. *Water Policy 17*(suppl.1), 58–88.

De Sá E Mendonça, B. C. & De Abreu, M. M. (2015). Integrated actions in the management of critical hydrologic events. *Water Policy 17*(suppl.1), 195–211.

Grabs, W. & Moser, H. (2015). Translating policies into actions: the case of the Elbe River. *Water Policy 17*(suppl.1), 114–132.

Grambow, M., Eichenseer, E., Overhoff, G., Hafner, T., Staton, K. & Schaipp, B. (2015). Lessons learned from the Big Floods, 1999–2013: The Bavarian Flood Action Programme 2020plus. *Water Policy 17*(suppl.1), 133–155.

Grobicki, A., MacLeod, F. & Pischke, F. (2015). Integrated policies and practices for flood and drought risk management. *Water Policy 17*(suppl.1), 180–194.

Kron, W. (2015). Flood disasters – a global perspective. *Water Policy 17*(suppl.1), 6–24.

Schultz van Haegen, M. & Wieriks, K. (2015). The Deltaplan revisited: changing perspectives in the Netherlands' flood risk reduction philosophy. *Water Policy 17*(suppl.1), 41–57.

Tachi, K. (2015). Lessons learned from mega-disasters and future policy development on water-related disaster management in Japan. *Water Policy 17*(suppl.1), 25–40.

Wieriks, K. & Vlaanderen, N. (2015). Water-related disaster risk reduction: time for preventive action! Position paper of the High-level Experts and Leaders Panel (HELP) on water and disasters. *Water Policy 17*(suppl.1), 212–219.

Flood disasters – a global perspective

Wolfgang Kron

Geo Risks Research, Munich Reinsurance Company, Koeniginstrasse 107, 80791 Munich, Germany
E-mail: wkron@munichre.com

Abstract

Various disasters in recent decades have confirmed that the risk from water-related events has been increasing significantly worldwide. Among those events are tsunamis, storm surges, river floods, flash floods, mass movements and droughts. The driving factors of this risk are the unabated increase in global population, the concentration of people in high-risk areas such as coasts, flood plains and hillsides, the rise in vulnerability of assets, infrastructure and social systems, and the consequences of environmental and climatic changes. Risk reduction requires general awareness at all levels of society and a partnership between the public sector, the people concerned and the insurance industry. Structural and nonstructural precautionary measures are always cheaper in the long run than paying losses. Overall economic consequences are significantly less severe in societies with a high insurance penetration.

Keywords: Disaster losses; Flood risk; Insurance; Risk management; Risk reduction; Water-related disasters

Introduction: recent water-related disasters

Water is responsible for most natural disaster losses in the world. Every year, various regions suffer from water-related disasters resulting from rainfall or snowmelt, storm surges, tsunamis, landslides or drought. In particular, 2013 was a year with some exceptional floods. The most severe in terms of loss of life and property occurred in the Philippines (storm surge during Typhoon Haiyan), central Europe and Uttarakhand (India), but there were also significant, and for the areas concerned disastrous, events in Colorado (USA), Canada, Mexico, Indonesia, Australia and southern Africa (South Africa/Mozambique/Zimbabwe) (Munich Re, 2014). In the first half of 2014, floods devastated vast areas in some Balkan countries, and a landslide in Afghanistan buried a village, causing about 400 fatalities. Droughts have been frequent and very costly in the United States in recent years, but have probably had more serious impacts on nations in East and Southeast Asia and in Africa's Sahel zone, especially Somalia and Sudan.

doi: 10.2166/wp.2015.001

The year 2011 saw the dramatic consequences of a tsunami and two extreme river flood events, one in Thailand and one along the Mississippi in the United States. The two river floods clearly showed how different the consequences can be depending on the different status of flood preparedness. The Thailand flood also revealed drastically the possible global effects of a local event.

Tables 1 and 2 display the greatest flood disasters in terms of monetary losses (>US$5 bn[1]) and those with the highest death tolls (>1,500) since the beginning of the century. Disasters with high monetary losses mainly occur in well-developed countries (Table 1), whereas high numbers of fatalities are usually related to events in poor regions (Table 2). But even though in poor countries natural disasters often do not produce high monetary loss figures in absolute terms – and sometimes not even high death tolls – they may still be severe and momentous for the country affected. On the other hand, events which cause comparably little physical damage – like the eruption of the Eyjafjallajökull volcano in Iceland in April 2010 – may produce costs running into several hundred million US dollars a day by interrupting private and business lives and the flow of goods.

Table 1 also reveals that the insured portion from the costliest events was much higher in developed than in poor countries. The insurance industry can assume a considerable part of the costs, thus relieving a nation's burden on the budget and economy. Therefore, disasters usually affect poor countries in a much more pronounced way than rich countries.

Disasters can take very different forms: in terms of scale (regional intensity or large-scale impact), high number of fatalities, huge monetary losses and severe impact on the local economy. Many disasters happen on coasts. Coasts, due to their concentration of people and assets in combination with a variety of hazards, are certainly the high-risk areas of the world (Kron, 2013). One third of the world's population lives here. There is no doubt that natural disasters, especially weather-related events, have been increasing dramatically in frequency and intensity.

The fact that large losses of life are less frequent nowadays than in the past testifies to the efficiency of modern technological achievements and efforts in disaster reduction, initiated for instance by the International Decade for Natural Disaster Reduction (IDNDR), which was proclaimed by the United Nations in the 1990s. However, Cyclone Nargis and the two great tsunamis showed it is by no means guaranteed that catastrophic death tolls are becoming more and more limited in water disasters.

Figure 1 shows (as percentage distributions for four types of water-related disasters and 'other' disasters in which water is not a significant factor) the number of loss events in the period 2000–2014, the fatalities they caused, and the overall and insured losses. In about half of the loss events (45%), fatalities (45%), overall losses (59%) and insured losses (53%), water plays a role. All percentages shown in Figure 1 are rather stable; they do not differ much from results obtained for the period 1980–2014.

It is normally not possible to attribute all losses in a disaster clearly to the different hazards. Tropical cyclones in particular almost always produce losses from wind and water (rainfall, storm surge), whose shares vary considerably from event to event. Attributing 50% to each of the two causes – wind and water – on average is probably not too far off from reality. In the category 'Other events', winter storms (through storm surges) and convective storms (through flash floods) also have water-related loss components, but their impact on losses is rather low, in the order of one percentage point. The 'Fatalities' ring chart in Figure 1 does not include the category 'Drought', which gives rise to victims indirectly through famine. The latter is generally related to poor societal conditions and therefore not

[1] bn = billion (10^9).

Table 1. Water-related disasters in the period 2000–2014 in which material losses of US$5 bn and more (original values) occurred.

Year	Country/region	Event/basin(s)/area	Overall losses due to water (US$ bn)	Insured losses due to water (US$ bn)	% insured (−)
2005	USA	Hurricane Katrina (Gulf Coast)	83^(2/3)^	41.5^(2/3)^	50
2011	Japan	Tsunami	55^(1/4)^	9^(1/4)^	16
2012	USA, Canada, Caribbean	Hurricane Sandy (Northeast)	46^(2/3)^	19.7^(2/3)^	43
2011	Thailand	Chao Phraya	43	16	37
2012	USA	Drought (Midwest)	20	12	60
2002	Central/Southern Europe	Elbe, Danube, Italy	16.5	3.4	21
2008	Caribbean, USA	Hurricane Ike	13^(1/3)^	6^(1/3)^	46
2013	Central Europe	Danube, Elbe	12.6	3.1	25
2002	USA	Drought (Great Plains)	10	2	20
2004	Indian Ocean	Tsunami	10	1	10
2008	USA	Midwest; Missouri	10	0.5	5
2010	Pakistan	Indus	9.5	0.1	1
2000	Italy, Switzerland	Southern Alps	8.5	0.48	6
2010	China	East, Southeast, South	8	0.15	2
2012	China	East, Northeast, Southeast	8	0.18	2
2004	Caribbean, USA	Hurricane Ivan	8^(1/3)^	5^(1/3)^	60
2003	China	Center, South, East, Northwest	7.9	–	–
2004	China	Southwest, Centre, Northwest	7.8	–	–
2005	Caribbean, Mexico, USA	Hurricane Wilma	7^(1/3)^	4^(1/3)^	57
2007	China	South, Southwest, East, Center	6.8	–	–
2001	USA	Tropical storm Allison (Houston)	6	3.6	60
2013	Canada	West (Calgary)	5.7	1.63	29
2004	Bangladesh, India, Nepal	Monsoon rains	5	–	–
2005	India	Monsoon flash flood (Mumbai)	5	0.77	15
2011	USA	Hurricane Irene (Northeast)	5^(1/2)^	3^(1/2)^	55
2013	Philippines	Typhoon Haiyan	5^(1/2)^	0.35^(1/2)^	7

Source: Munich Re NatCatSERVICE 2015.
(1/4)/(1/3)/(1/2)/(2/3) The loss figure shows the – roughly estimated – losses attributed to flood (one quarter/one third/half/two thirds of the overall/insured losses); the remainder is attributed to windstorm or to ground shaking during earthquake. Example: the overall loss of Typhoon Haiyan was US$10 bn, ½ of it (US$5 bn) was due to water.

primarily a consequence of a natural phenomenon. As tsunamis are always secondary phenomena of 'Other events', they are not shown separately in the 'Number of loss events' ring chart.

The number of inland flood loss events captured in databases shows a distinct upward trend. Figure 2 displays those events exceeding a threshold of US$50 million in today's value since 1980. If all events,

Table 2. Flood disasters in the period 2000–2014 in which more than 1,500 people died.

Year	Region	Event	Deaths*
2004	Indian Ocean (12 countries)	Tsunami	220,000
2008	Myanmar	Cyclone Nargis	140,000
2011	Japan	Tsunami	15,880
2013	Philippines	Typhoon Haiyan	6,334
2013	India	Flash floods	5,500
2007	Bangladesh	Cyclone Sidr	3,295
2004	India, Bangladesh, Nepal	Floods	2,200
2007	Bangladesh, India, Nepal	Floods	2,030
2004	Haiti, Dominican Republic	Floods	2,000
2004	Caribbean, USA	Hurricane Jeanne, floods	2,000
2010	Pakistan	Floods	1,760

Source: Munich Re NatCatSERVICE 2015.

*Death figures include all causes (such as earthquake, windstorm, landslides, etc.) not only flood; those missing are not included.

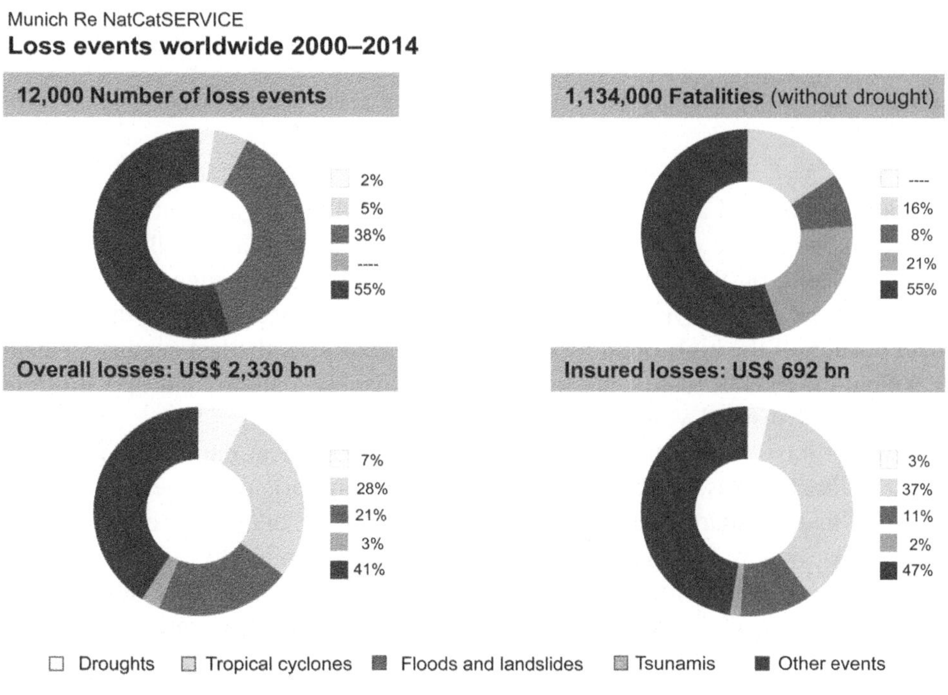

Fig. 1. Percentage distributions of number of events, fatalities, overall losses and insured losses for water-related loss events and other types of natural disasters. Period: 2000–2014. Losses are given in inflation-adjusted values of 2014, with calculations based on the Consumer Price Index of each country. *Source:* Munich Re NatCatSERVICE 2015.

even small ones, were considered, a strong reporting bias would govern the jagged line of the number of events, whereas the bars for the losses would not be influenced significantly. Several decades ago, small events that happened in remote areas did not always find their way into global databases and

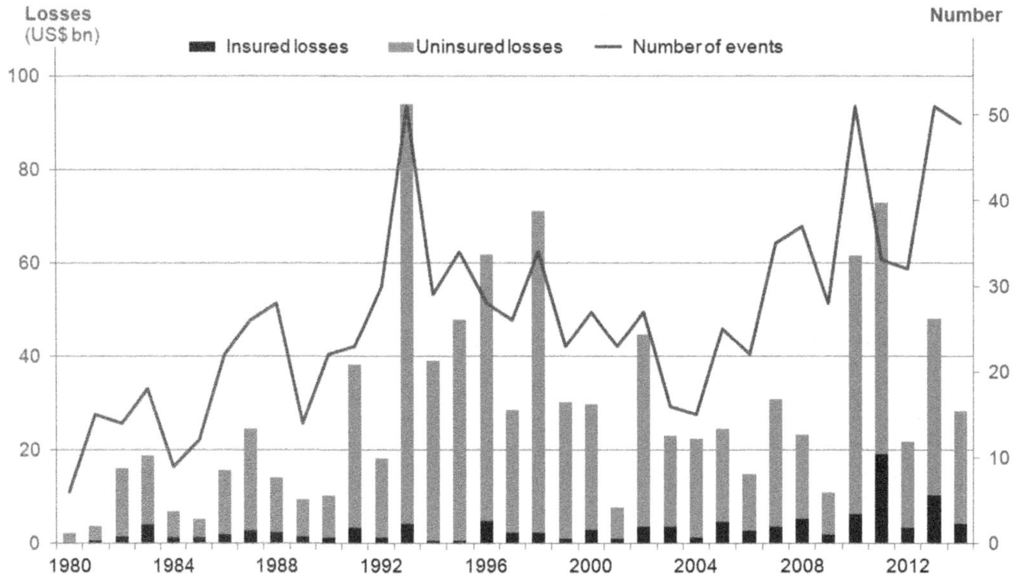

Fig. 2. Inflation-adjusted (to 2014 values) overall and insured annual inland flood losses (bars) and number of flood events per year (jagged line) from 1980 to 2014, derived from all events with losses exceeding US$50 million (in 2014 values). Only floods that are not associated with named tropical cyclones are included in the analysis (Kron *et al.*, 2012). Source: Munich Re NatCatSERVICE 2015.

are therefore underreported and also did not produce considerable loss amounts. The rise in the jagged line in Figure 2 reflects, on the one hand, the increase in population and built-up areas, especially at flood-prone locations and, on the other, is most likely also a climate change signal.

A comparable increase in the overall losses – at least since the beginning of the 1990s – cannot be stated. The main portion of the high flood losses in the last decade of the last century comes from events in China. Other than the overall losses, the insured losses display a significant upward trend caused by the increasing portion of insured values.

Recent large events have shown that losses resulting from physical damage are no longer the only significant ones. Indirect losses, such as business interruption (bi), contingent business interruption (cbi) or the loss of market share that a company suffers due to being out of business for a while, have assumed completely new dimensions. These losses are not limited to the area affected by the disaster but may occur anywhere in the world, even far away from the site of a flood. For instance, the floods in Thailand caused a globally felt shortage in hard disk drives when a plant had to be closed. One quarter of all hard disk drives are manufactured in Thailand. While this has the ('positive') effect that the area hit does not have to carry the entire loss alone, it makes disasters less 'calculable' for industry, especially for the insurance industry.

Types of water-related disasters

Flood events can be divided into four main types: river floods, flash floods, storm surges and tsunamis. Additionally, there are phenomena such as lake flooding, groundwater flooding and waterlogging, dam break and glacier lake outburst floods, backup floods, seiches, etc. (Kron, 2012). Most tsunamis are generated by geophysical phenomena. Besides floods, droughts and certain pollution disasters are also water-related, the latter being rarely of natural origin though. The three most severe 2013 events represented the three main hydro-meteorological causes.

River flood

The 2013 flood in central Europe was a typical river flood. River floods (or basin-wide floods) are the result of copious rainfall (or snowmelt) continuing for a period of days or weeks over a large area during which the basin's (soil's) retention capacity is depleted. The drainage system of a catchment area conveys the rainwater directly to the main river(s). This type of flood does not occur suddenly. A flood wave is built up gradually, though sometimes in a relatively short period. As it propagates downstream, it can affect many reaches along the course of the river. The affected area may cover entire river basins and thousands of square kilometers. Water levels often remain high for a long time and interfere with the lives of people, not only by damaging property but also by preventing them from living their normal lives. In river floods, inundation always emanates from the river channel, and the size of the flooded area is usually a function of peak discharge and flood volume. The sequence of affected areas is therefore always the same, and zones with a different hazard (as regards flood frequency, but sometimes also flood intensity) can be identified.

Flood control systems can have a significant impact on the extent of flooding. Precautionary measures are effective, as the source of flooding – the river course – is known. Dikes, retention basins and diversion channels can be constructed. Early-warnings may be issued allowing for preparedness measures and evacuation – given that forecast and observation systems, forecast models and communication lines are in place. An adverse effect of highly developed flood control and preparedness is the 'feeling of security' that is created in the people protected. They tend to forget about or ignore the residual flood risk if nothing happens for a while and increase the risk by accumulating assets.

Flash flood

Flash floods (or off-plain floods) can happen practically anywhere. These floods are produced by intense rainfall, usually of short duration. They often occur over a very small area in conjunction with thunderstorms or over a large area during tropical storms. Flash floods have a sudden onset and last from a few hours to perhaps a day or two. This surprise effect is the reason why flash floods pose such a danger for people and are responsible for many lives lost each year. In sloped terrain, the water flows at high speed and has an incredible potential for destruction, whereas in flat areas the water cannot flow off fast enough due to the lack of a sufficient gradient and accumulates on the surface or in depressions. Flows in affected watercourses grow rapidly and flood waves can gush downhill in next to no time and forge into areas where it may not even have rained at all. Mechanical forces, erosive capacity and transport of solids are critical factors associated with this type of flood. Sometimes flash floods mark the beginning of a major river flood, but often they are separate, individual events of only local significance, scattered randomly in space and time.

It is almost impossible to forecast flash floods adequately – only the general threat due to a certain weather situation can be stated. In most cases it is only a question of minutes, thus more or less ruling out short-term loss reduction measures. The floods in India in 2013 were of this type.

Storm surge

Storm surges can occur on the coast and along the shores of large lakes. Since they are created by wind, they are of meteorological rather than hydrological origin. They have extremely high loss potentials and have caused hundreds of thousands of fatalities, even in the recent past. In Bangladesh, death tolls of 300,000 and 139,000 were reported in 1970 and 1991, respectively, and 140,000 died in Myanmar during cyclone Nargis in 2008. Typhoon Haiyan claimed more than 6,000 lives in the Philippines in November 2013. In the past decade, storm surges induced by hurricanes Katrina (2005), Ike (2008) and Sandy (2012) have caused huge physical damage in North America.

In Europe, the United Kingdom, the Netherlands and Germany have been able to reduce the risk thanks to immense efforts in the past decades. Major improvements in sea defences and, in particular, the enhancement of forecasting and early-warning facilities in recent years have led to great storm surge disasters becoming less common. Nevertheless, storm surges still represent an immense loss potential in what is a relatively limited strip of land on the coast.

In many Asian countries, physical protection against storm surges is not available. This is not only related to the enormous financial means necessary to construct them, but also to the physical conditions of the coast. Huge delta coasts fissured by numerous outlets and consisting of hundreds of kilometers of irregular coastline cannot easily be protected by dikes.

The accelerating rise in sea levels that is to be expected will aggravate the risk of storm surge and coastal erosion all around the globe – and this will be one of the most detrimental effects of global warming.

Tsunami

Tsunamis are gravity waves, generated by large volumes of water being displaced by large earthquakes, submarine or onshore landslides, or volcanic eruptions. With a wave height of a few decimeters and a wave length of up to several hundred kilometers, they travel through the open sea at the speed of a passenger jet without losing much of their energy. As they approach the coast, their speed decreases but at the same time their height increases – sometimes drastically. Wave crests more than 10 m high have been observed. These waves may reach run-up heights exceeding 30 m on the shore and destroy everything in their way. Very often the water is seen to retreat from the shore before a tsunami reaches the coastline, giving people a chance to escape. The two disasters around the Indian Ocean in December 2004 and on central Japan's Honshu coast in March 2011 dramatically demonstrated the destructive power of this natural phenomenon, against which technical protection remains very limited. Education and knowledge about adequate behaviour as well as early-warning, however, are very effective in saving lives. Probably most of the 220,000 people who perished in the 2004 tsunami would still be alive today if they had read the signs properly or been adequately warned.

Mass movement

Whenever abundant water meets sloped terrain, mass movements can be triggered. Moving solids are much more destructive and powerful than water alone. Water in debris flows and lahars is the main driving factor, whereby it serves primarily as a lubricant in landslides that reduces friction and leads to instability of a slope. Although highly devastating – four landslides in 2014 wiped out a village in

Afghanistan killing 400, hit two places in India and Nepal claiming more than 150 victims each, and left 43 dead in Oregon/USA – single mass movements usually have only a local impact. However, sometimes thousands of mass movements occur at one time, such as in 1999 when several thousand people perished following landslides and debris flows in northern Venezuela.

Drought

Lack of water, unlike disasters involving too much water, does not immediately become critical. Droughts develop over time. They can directly lead to agricultural losses by preventing crops from growing or by leading to starvation of livestock. Loss figures can be enormous (cf. Table 1), but usually only become truly catastrophic if the affected society does not have the organizational means to cope with the consequent food shortages and widespread starvation which may result in tens of thousands of deaths. It is often very difficult to recognize the beginning of a drought-related crisis. Defence measures, as a rule, come too late. With the world now more interconnected, the large human losses in the 1970s (Sahel zone, India) are a thing of the past, but nevertheless the hazard is still able to cause major disasters as the impacted area is usually large.

Reasons for increasing losses

Disaster losses are driven by the increase in global population and wealth, the settlement and industrialization of regions with high exposure levels, and the fact that modern technologies are highly susceptible to external disturbances. Due to its complexity, international trade is even more susceptible. Changing environmental conditions, in particular climate change, and the lack of adequate risk perception are additional features. In contrast to loss reduction measures against other types of natural hazards, those against floods can be quite effective. This may be one reason why no distinct upward trend in the monetary losses can be identified (cf. Figure 2).

People – land use – risk awareness

At the moment, there are about 7.3 billion people on earth, up from three billion in 1960 and heading for nine billion in the 2050s. Adequate and safe areas for them to live are limited by natural circumstances, but also by economic needs. While in the past settlement areas could be chosen because of the shelter they offered against adverse natural conditions, nowadays any available piece of land has to be used. Furthermore, new residents are often unfamiliar with the local hazard situation or are lulled into a sense of security by trusting in technology's ability to control the forces of nature.

Coastal areas, in particular, have been very attractive to people. Today, one tenth of the world's population lives in low-lying areas along the coast, less than 10 m above mean sea level (Lichter *et al.*, 2010), one third within 100 km of a coastline (Small & Nicholls, 2003), and about half within 200 km (Creel, 2003). Fifteen of the 20 largest cities in the world are located on the coast. The trend of migration toward coastal areas is likely to persist. According to an Organization for Economic Co-operation and Development study (Nicholls *et al.*, 2008), 113 million people will live in the flood-prone neighbourhoods (100-year flood zone) of the 20 most populated coastal cities in 2070, an almost five-fold increase on today's number. The same study predicts that flood-prone assets in the 20 cities with the highest concentration of flood-exposed values will increase from the current US$2.2 bn to about US$27 bn. In addition to permanent residents, millions of tourists choose coastal regions as their holiday destination.

In the interior of countries, river plains are – if one disregards the flood hazard – also well suited for development, and frequently used for this purpose. While flood control measures prevent frequent losses and inconvenience, this effect is counterbalanced by the feeling of security it creates, leading people to expose more and more objects of increasing value to the risk of flood. This sense of security is transmitted not only by dikes and embankments, early-warning systems and the availability of disaster-relief organizations, but also by the intentional or unintentional transmission of false information and by local authorities or groups with a vested interest (e.g. the tourist trade) playing down the risk. While this is a worldwide problem, there are some striking examples such as areas along the Yangtze River in China that were devoted and designed for flood retention in the 1950s, but can no longer be used for this very purpose as several hundred thousand people live there today. One example is Jinjiang polder on the Yangtze River. It would be impossible to evacuate the 600,000 residents in time if a major flood occurred. And they would face an inundation depth of 6–9 m.

Complexity – wealth – susceptibility

The two big disasters in Thailand and Japan in 2011 showed how badly we depend – even on a global scale – on functioning supply chains (Munich Re, 2012a). Locally this is also true for power supply and telecommunication networks. Given a complex infrastructure, a failure in one place may cause a domino effect that brings a whole system to a standstill.

Wealth has increased in practically all regions of the world. At developed locations, even in poor countries, the potential for destruction is high, as more and more expensive items can be found inside buildings than was the case in the past. In general, modern equipment is highly vulnerable. Almost everything contains electric or electronic components, and these items often suffer severely when exposed to water and even humid, salty air. Whereas years ago water-damaged items had simply been dried and re-used, they are now discarded.

Urban concentration and environmental changes

It is not only the development of hazard-prone regions, but additionally the concentration of commercial and industrial centres which attracts people. More than 50% of the world's population lives in urban areas compared to just 30% in 1950 – and the percentage is still increasing. Half a century ago, there were eight metropolitan areas in the world with a population exceeding five million. Today the number of megacities has grown to over 70 (Demographia, 2014). It is obvious that the chance of a severe natural event hitting one of these high-concentration regions is steadily growing.

Colonization of land always has the consequence of changing the environment. Forests are removed, rivers have to be tamed or diverted, and the local micro-climate is possibly altered. Most of these measures improve the situation in the short term but worsen it in the long term. One fairly common consequence of urban growth is subsidence, caused for instance by groundwater extraction or sometimes simply by the weight of buildings erected on relatively soft and unconsolidated coastal soil. A few decimeters of subsidence can create huge problems in already low-lying port cities and deltas, some areas suddenly finding themselves below a given flood level or even below the local sea level.

Climate change

The scientific facts are clear: the global average temperature close to the ground has risen by 0.85 °C since 1880. Climate change is taking place and it is mainly caused by human activities. The Fifth Assessment Report of the Intergovernmental Panel on Climate Change (IPCC, 2013) confirmed earlier findings in this regard and clarified in many fields the statistical confidence of results. With respect to water-related disasters, we are headed for higher sea levels along most coasts, more and more intense extreme precipitation events, and more and prolonged dry periods in many regions. In some places, storminess will increase leading to higher waves and elevated storm surge levels.

From hazard to risk

It is important to understand the circumstances under which natural disasters happen. Nature alone does not produce disasters, it only produces extreme events. A natural disaster happens if people and/or their possessions are affected so severely that a society's life is disrupted. A well-prepared society is not likely to experience a disaster as easily as one where many aspects of preparedness are missing, from education and knowledge to building codes, and from functioning governance to availability of financial means, making it vulnerable to impacts from nature. Disasters are hence not only products of chance but also the outcome of interaction between political, financial, social, technical and natural circumstances.

Whether a location is risky depends on: (a) the likelihood that a natural event may occur; (b) the presence of values; and (c) their vulnerability (Kron, 2005). Where there are no people or values that can be affected by a natural phenomenon, there is no risk. In a simplified, but widely accepted way (IPCC, 2013) the term 'risk' can be defined as

$$\text{Risk} = \text{hazard} \times \text{values at risk} \times \text{vulnerability}$$

Hazard is the threatening natural event, including its probability of occurrence; values at risk are the buildings/items/humans that are present at the location involved; and vulnerability denotes the lack of resistance to damaging/destructive forces. Vulnerability can refer to human health (human vulnerability), structural integrity (physical vulnerability) or personal wealth (financial vulnerability). Insurance's contribution to risk control addresses the last of these factors. In its most simple form, risk is computed by multiplying these three components. Values at risk and vulnerability are sometimes combined to form 'consequences'. Thus, risk can also be written as

$$\text{Risk} = \text{hazard} \times \text{consequences}$$

The overall risk is determined by computing the integral over all possible threatening events (intensities and frequencies) with their respective consequences (associated losses). Hence, risk is identical to the (expected) average annual loss.

All three components of risk have been and still are increasing unabated. On the one hand, rising sea levels, increased tropical cyclone intensities (wind and rain) and unprecedented impacts of floods and, on the other hand, megacities with exploding populations and industrial development are making many regions ever riskier, in particular those on coasts and in flood plains.

A partnership for risk reduction

Risk and loss minimization call for an integrated course of action. The risk must be carried on several shoulders: the state, the people and enterprises affected, and the financial sector, in particular the insurance industry. Only when they all cooperate with each other in a finely tuned relationship, in the spirit of a risk partnership, can disaster prevention really be effective.

The job of public authorities (i.e. the state or the government) is primarily to reduce the underlying risk to society as a whole. They provide access to observation and early-warning systems, build river dikes and sea defences, determine the framework for the use of exposed areas by enacting statutory provisions, and prepare emergency plans, including programs to alleviate recovery (temporary housing, financial assistance, tax relief, etc.). In some countries, insurance programs are state-run. Unlike in the case of earthquake and windstorm, where homeowners themselves are responsible for ensuring their houses are properly protected, the responsibility for protection from flooding is largely shifted to public authorities.

Those immediately affected (individuals, companies, communities) have a great potential for loss reduction. The crucial point is whether they keep their risk awareness alive. Even those people who do not neglect the danger of a natural peril from the very beginning often quickly forget about it, especially if nothing happens for some time. They rely on technological protection systems and at the same time make their property more and more valuable by adding additional items that are often susceptible to damage. These people must be informed and educated to build in an appropriate manner, control the exposure of their values, and be ready to take action in an emergency. This includes preparing for catastrophic losses by taking financial precautions, e.g. buying insurance.

The true task of the insurance sector is to compensate financial losses that would have a substantial impact on the insured or even bring about their ruin. They carry the financial risk from events that have such a low probability that they cannot be considered foreseeable. Insurance redistributes the burden borne by individuals among the entire community insured, which is ideally composed in such a way that they all have a chance of being affected – even if the degrees of probability differ.

Unfortunately, the penetration of insurance, i.e. the percentage of the population that has insurance cover, is strongly correlated with the wealth of a society. The times when governments alone took full care of their people are over. Today, owners are more and more responsible for their belongings themselves and are required to protect themselves by buying insurance. From this, it follows that poor people who have higher priorities in their daily lives with survival struggles (such as food, health, shelter, etc.) than preparing for a future catastrophe event with some probability are rarely, if ever, insured.

Flood insurance

Flood insurance has become a top issue in many countries in the aftermath of the large flood disasters that have occurred all over the world in recent years. At first sight, flood insurance that is obligatory for everybody seems the best solution to bypass the necessity of spending huge amounts of public money on financial relief and reconstruction of damaged private property. The problem is that the potential insured are generally not happy about subsidizing those who are exposed to a risk much higher than their own. On the other hand, the loss-bearing capacity of the insurance industry and the financial sector is limited. The state must always serve as a reinsurer of the last resort, stepping in when losses exceed this capacity. However, governments are reluctant to guarantee the assumption of potential record losses. This would require them to build up funds, with money that would then not be available for other purposes such as healthcare, pensions and infrastructure projects.

A purely voluntary insurance scheme would fail to reach large parts of society due to adverse selection or lack of risk awareness. These people would rely on and seek help from the state in the event of a disaster. A system should be set up that finds a compromise between the extremes of fully obligatory and fully voluntary covers: it must be based on risk-adequate premiums that are cross-subsidized to some extent to make them affordable but are nevertheless acceptable to the vast majority.

The state should also restrict post-disaster-relief payments to those who are not insurable but have otherwise contributed to their own protection. At the lower end of the scale, it must be acknowledged that insurance should not be simply a band-aid that takes care of any little financial injury but is meant for substantial and catastrophic incidents that could ruin an insured individual.

As a rule, insurance companies react very quickly to losses from natural events and reimburse their clients. Apart from the benefits of the very helpful financial support they provide, the psychological effects of prompt assistance are very important for the people concerned. They do not feel left alone in an extremely difficult personal situation. This problem occurs if governmental financial aid is promised but the payments are delayed for a long time for administrative reasons. Also, the actual payments seldom match the amounts promised by politicians immediately after a disaster. For insurance companies, which do not need to go through long political decision-making processes and can react immediately, quick help has its advantages too. Their mottoes are 'the faster, the cheaper' and 'good loss adjustment is the best form of promotion'.

Hazard zoning and premiums

Calculating the price of insurance is – in theory – very simple. The long-term premium payments by the insured must basically be equal to the long-term compensation payments to the insured (leaving alone other aspects such as interest and administration loadings). A loss that occurs once in 100 years on average therefore requires a premium of 1% of this loss per year. These are statistical considerations: some insured do not suffer a loss at all, while others suffer it several times in that period. Hence, a portfolio is needed that allows averaging over time and space by insuring a large number of individuals. The individual premium should represent the individual risk. As this is not feasible for each and every small object to be insured, classification into zones with a similar hazard level is performed, with a lumped premium as a percentage of the insured value.

Insurance contracts are based on the status quo. If the hazard is high and the precautions taken minimal, the premium charged will be high (if the object is insurable at all); if the hazard is low or protection standards are high, premiums can be low. In practice, insurers often recommend (or require) precautionary measures (structural, emergency plans, etc.) to reduce the risk and also to make it more calculable.

Pricing must also take into consideration the cost of reserves being kept at hand for large losses. The probability of a single home (e.g. in the Netherlands just behind the sea dike) being flooded is very small ($<10^{-3}$), but if that occurs, thousands of other houses will be damaged at the same time and the overall loss will be enormous. Maintaining reserves for such a contingency costs a multiple of the amount reflecting the individual risk; it thus constitutes another component of the premium. The required annual premium – in the example – is therefore not around 0.1% but maybe 1% of the insured value, which makes the insurance cover practically unaffordable.

Today, insurance for floods – and other natural perils – is widespread in only a few countries. The current slow increase in the insurance density is unlikely to change dramatically in the near future, because both sides – customers and insurers – are not really promoting such a development. On the one hand, customers' demands are impaired by a growing number of people with a lack of financial resources and, on the other, the willingness of insurance companies to cover certain risks is very limited because of their magnitude, the great amount of uncertainty involved or because of accumulation control considerations (e.g. insurance for storm surges). Wherever natural disaster insurance is possible, available and in operation, however, it will help – by spreading the risk throughout the world via the reinsurance sector – to reduce the vulnerability of society in terms of its exposure to natural hazards.

The role of reinsurance

In the same way that private individuals/entities do (by buying insurance), insurance companies also try to avoid volatility in their payments. Natural perils insurance is highly volatile. Large single losses (from one event) can be reduced by transferring part of the risk to the reinsurance sector. While most insurers concentrate their business on a particular country or region, reinsurance companies do business worldwide, thus relieving the local insurance market and possibly even preventing its collapse. How effectively this idea of transferring local losses via the reinsurance sector to a worldwide system works is shown by the example of Hurricane Gilbert, which struck the Caribbean in 1988. Jamaica in particular suffered great losses; its economy was hit by losses amounting to about US$1 bn, of which 70% was insured. This US$700 million would have destroyed the Jamaican insurance industry completely. It survived because nearly 99% or US$690 million was reinsured and was therefore paid by the world's reinsurance industry. For the local companies a mere US$10 million obligation remained.

Typhoon Haiyan in the Philippines (2013) yielded a similar picture. Although losses were still not finally settled at the end of 2014 and figures are therefore still uncertain, the reinsured percentage for the event is most likely to be more than 90%. A reinsurance rate of more than 95% is typical for developing countries. In developed countries, reinsurance rates usually range between 50 and 90%, depending on the strength of the primary insurance companies in the region. Since reinsurance costs money, large primary companies tend to keep a larger portion of the risk themselves.

It is not easy to give examples of the reinsured portions of losses for large natural disasters. For example, about US$10 bn of the total insured loss of US$16 bn for the Thailand floods in 2011 were industrial losses from Japanese companies, insured in Japan and reinsured there with a rate of about 80%. Some US$5 bn of the remaining US$6 bn was reinsured via the domestic Thai market. This means that not all reinsured losses enter the international, global market directly; there are also domestic reinsurers (e.g. Thai Re in the case in question) that may assume quite a large chunk of the losses, which thus remain in the country. Only part of these losses is passed on to the international reinsurance market by purchasing retrocession. For the Thailand floods, the estimated amount of internationally reinsured losses was eventually US$13 bn, or 81% of the US$16 bn total insurance burden.

This system of insurance, reinsurance, retrocession combined with complex, often layered contracts involving co-insurance, deductibles and limits shared between several/many reinsurers makes it difficult and often almost impossible to specify even rough figures of reinsurance shares – especially given the confidentiality of the business-related figures of the companies involved. In general, reinsurance shares of losses are higher for large loss events than for less costly ones.

Insurance and especially reinsurance companies must be prepared to pay large amounts of money after major events. The company must not be threatened in its existence even by such enormous amounts.

Money for payments must be made available very quickly and cannot be placed in long-term – and thus profitable – investments. This means volatility is expensive. Assessing the probable maximum (accumulation) loss from an extreme event and holding adequate reserves are crucial to an insurer's economic survival. The companies therefore use models simulating a large number of – stochastically generated – events to obtain a probability distribution of potential losses. From this distribution, the expected loss for a given return period (chosen according to a company's business policy), can be read off.

While, in the past, the major reinsurance companies developed and provided accumulation loss models, there is a considerable market for these models today. Due to the rising number of physical parameters and processes considered in the calculations, the models' complexities have become enormous and the industry has to rely on only a few different providers per hazard and region. The large reinsurance companies maintain and improve their own models too, but often use market models additionally for a second opinion.

Reinsurance companies, due to their worldwide activities, are among the best sources for natural disaster statistics (Kron *et al.*, 2012). They systematically register disaster-relevant data in their databases as a basis for numerous analyses. The databases focus on three aspects: the number of people affected (fatalities, injured, homeless), the overall economic damage – i.e. direct losses – to the country affected, and the losses covered by the insurance industry (Wirtz *et al.*, 2012). Indirect losses and other consequences (e.g. jobs, impeded domestic growth, loss of market shares due to business interruption, etc.) are not considered.

Munich Re's Geo Risks Research department has been systematically collecting information on natural disasters for more than 40 years. Its NatCatSERVICE (NCS) database, the world's largest on losses from natural disasters, contains more than 35,000 entries. In recent years, many other institutions have recognized the importance of data on disasters losses and started to construct databases.

The NCS data and the data analyses are not only used for determining reinsurance premiums but – although not generally accessible to the public – are also made available to governments and non-governmental organizations to assist them in better planning and developing prevention measures against natural disasters. Close contact with scientific institutions worldwide (including participation in the work for the Fifth IPCC Assessment Report 2014), research cooperation projects and involvement in scientific bodies/boards/committees are maintained, and assistance is offered for political and administrative decisions that may be based on the data. In general, reinsurers' expertise on risk assessment and risk reduction is welcomed, as is their experience with disasters – their origin and evolution, the processes involved, and how to cope with them.

The insurance industry plays an important role in raising awareness and coping with natural hazards: it quantifies risk by means of adequate premiums and thus makes risks transparent. Therefore, it creates incentives for reasonable behaviour and prevention, and so reduces the losses for society. It also has tremendous potential for promoting climate protection and climate change adaptation, and thus positively influencing future losses, by taking account of such issues in its products, investments, sponsorship activities and communications. Companies perform educational and public relations services, e.g. by publishing brochures in which they draw attention to hazards and explain ways of dealing with them (e.g. Munich Re, 2012b, 2013).

Economic consequences of natural disasters

Impacts from water disasters are not as devastating to rich societies as to those in less developed parts of the world; there, whole countries are sometimes thrown back in their development by years. On the other hand, rich states have a significant financial burden not only from disaster losses, but also from the costly precautionary measures that citizens demand from their governments to protect themselves and their properties.

It is often the countries with low per-capita income which must overcome large economic losses from natural disasters relative to their overall economic strength. These economies usually lack the financial resources needed for both prevention of disasters and disaster-relief. The floods in Thailand in 2011 cost US$43 bn which was 12% of the country's gross domestic product (GDP). Hundreds of thousands of homes, a vast amount of agricultural land and important industrial areas were flooded, and 65 of Thailand's 77 provinces were affected. Indirect losses from natural disasters such as delays or interruptions to production must be added to these figures. The result was that during the most serious phase of the floods in the fourth quarter of 2011, Thailand's GDP shrank by 2.5% compared with the previous quarter. For the Philippines, the World Bank has calculated that indirect losses from typhoons reduce GDP growth every year by 0.8 percentage points.

Other important macroeconomic parameters such as government debt or external trade also reflect the negative indirect effects. In Chile, the balance of trade collapsed and debt rose by around 70% in 2010 after the earthquake. A World Bank study found a statistically significant deviation of up to 30% from the historical trend of the per-capita government debt in emerging economies in the five years after 'major' natural disasters (Melecky & Raddatz, 2011).

Natural disasters can have a positive effect on an economy because reconstruction acts as an economic stimulus. Additionally, destroyed old assets are replaced by new production facilities and infrastructure which are generally of better quality. After the tsunami in 2004, the economy of the Maldives shrank by 8.7% in 2005 and then grew again by 19.6% in 2006 – stronger than in any of the previous 20 years. It must be noted though that the above-average growth was due, at least in part, to the low comparison base caused by the disaster. In a similar way, Thailand experienced a strong upswing in the year after the floods.

Nevertheless, the indirect positive effects on GDP cannot offset the indirect losses of all countries and natural disasters. Considering natural disasters with over 100 dead or US$250 million direct losses (after adjustment for inflation) von Peter *et al.* (2012) found with significant statistical evidence that GDP was almost 4% lower after five years than it had been without disasters. Overall losses (direct and indirect) related to their GDP were larger on average in developing and emerging economies than in wealthy industrialized countries.

However, GDP alone is not an ideal measure for the recovery of a country. Disaster often has an effect on national debt. It is therefore necessary to include national debt and 'destroyed capital' when assessing the impact on a country. In poor countries, the resilience effect – i.e. bouncing back to the previous value – is very much delayed, in some cases by decades.

Insurance has a positive effect: given two countries with identical per-capita income, the country with higher insurance cover will be better able to withstand natural disasters. This applies to the economy as a whole as well as to the insured individuals. Similarly, the higher the insurance cover against natural disasters, the lower the anticipated government debt, external trade deficit and macroeconomic impact.

What can and should be done?

Half a century ago people started to believe that humankind was well on the way toward controlling nature. Understanding the genetics of natural processes combined with technical development seemed to be the basis for gradual but continuous improvement in dealing with natural events. The catastrophic events that fill the annual reviews prove that we are still far away from having control, anywhere in the world. The UN IDNDR from 1990 to 1999 was certainly an important period in which significant progress was achieved in many fields related to disaster risk reduction. Through this initiative, but also by the efforts of countries, regions and cities, measures were undertaken to respond to the increasing threat of natural hazards.

The effect of protection and preparedness measures is clearly visible in many examples. In Bangladesh, fatalities from tropical cyclones were enormously reduced through early-warning and flood shelters. Cyclone Sidr in 2007, which was comparable to the 1970 and 1991 killer storms that claimed 300,000 and 139,000 lives, respectively, was responsible for 'only' 3,295 deaths, although Sidr was more powerful than the 1970 storm that hit the same area.

The city of Hamburg (Germany) has invested €2.4 bn (US$3.2 bn, in 2014 values) in flood protection since the 1962 storm surge disaster. At least five events with a higher water level than in 1962 have occurred since then without producing significant losses. It is estimated that further storm surge losses of at least €20 bn have been avoided thanks to the measures taken (Kron, 2014). In a similar way, the Netherlands have also prevented storm surge disasters since the catastrophe of 1953 thanks to their Delta Program (Wieriks & Schultz van Haegen, 2015).

In 2011, a centennial flood rolled down the Mississippi. The losses incurred were in the order of US$5 bn, but would have been tens of billions if the Mississippi River and Tributaries flood control project had not been carried out in the past decades (Delli Priscoli & Stakhiv, 2015). Also, China made immense efforts to improve its flood situation, particularly in central China, after the devastating flood season of 1998 when the Yangtze and Songhua Rivers caused almost US$30 bn of damage. Although billion-dollar losses in China still occur frequently, the middle and lower course of the Yangtze has been spared major problems in the 21st century despite some critical hydrological events.

A lot can and must be done in different scales and on different levels. Disaster prevention and risk reduction start with protecting the global climate from becoming more and more threatening, and range via forecast, warning and technical control systems to the individual's (person, company) behaviour and provisions to make sure she/he/it will not be ruined by an extreme event. While there is no discussion that loss of life must be prevented by all means, the costs of the efforts to prevent monetary losses should not be completely out of proportion to the value of the protected items.

While flood insurance penetration is low in developed countries because people lack risk awareness, people and businesses in poor countries often simply cannot afford to pay insurance premiums, if the cover is available at all. In the aftermath of floods, it is most often the poor that have to rebuild their lives without the financial cushion of insurance. Effective and economical solutions have therefore to be found for emerging countries. These countries – especially in Africa or South and Southeast Asia – are often faced with a higher chance of natural disasters. At the same time, the level of insurance density is generally very low, as these markets are less industrialized and people have less available income to insure themselves against existential risks.

Insuring the poor

People in developing countries need help to adapt to increasing losses caused by floods and other extreme weather events. One attempt to provide support for climate change-related risks is the

Munich Climate Insurance Initiative (MCII), which suggests risk management systems. Subsidies of insurance premiums by the industrialized countries predominantly causing climate change could be part of such solutions. Some of MCII's proposals have already found their way into the negotiation texts at the world's climate summits.

An MCII weather index insurance was launched in the Caribbean in mid-2013. For the islands of Saint Lucia, Jamaica and Grenada a 'Livelihood Protection Policy', developed together with the Caribbean Catastrophe Risk Insurance Facility and a micro-insurance consultant was introduced. In this product, insurance cover is triggered if specific meteorological parameters (e.g. rainfall intensity) exceed defined limits. This means prompt payouts following weather events without complex claims settlement.

Micro-insurance schemes are a key to helping people in emerging countries manage their risk better and maintain their standard of living. Micro-insurance is specially designed for low-income people and covers them against specific perils in exchange for regular premium payments proportionate to the likelihood and cost of the risk involved. Generally, micro-insurance is for people ignored by mainstream commercial and social insurance schemes – people who do not have access to regular products. Basically, it works like any other insurance scheme, but there are some aspects that make it different. Firstly, it does not cover a single client under one contract, but rather thousands of clients. Secondly, micro-insurance requires an intermediary between the client and the insurance company, a (local) non-governmental organization, for instance, or a rural bank that can handle the distribution and administration. Furthermore, it involves a variety of problems such as low premiums and high transaction costs per client, lack of infrastructure, lack of insurance knowledge, insurance illiteracy (clients do not understand the concept of insurance), low and irregular income, and lack of data. Therefore, the first and most important prerequisite for an effective micro-insurance scheme is raising awareness and educating the people concerned.

Micro-insurance covers for risks related to natural disasters are still hardly available. However, as fewer than 100 million people (or 3%) of the world's poor have access to insurance, the micro-insurance market has huge development potential. One long-term objective, of course, is also to help people develop a stable life situation that allows them to choose regular insurance products and thus become normal clients of the insurance industry.

The role of the global financial markets

The availability of huge amounts of money in the international financial markets, the potential of record losses from single natural disasters and the need for special products have led to the emergence of new financial instruments which supplement the classic distribution of risk via the reinsurance market: these are called catastrophe bonds (cat bonds). Through a cat bond, a specified risk (e.g. losses from a hurricane in Florida) is transferred from a risk carrier (sponsor) to investors. The sponsor is usually a member of the (re)insurance industry, but may also be a large company (such as a national railroad company). The investor buys a share of the bond and earns interest on the purchased notes. If the catastrophic event for which the bond is issued does not occur during its lifetime, the invested capital is paid back to the investor at maturity, together with the interest. If it does occur, however, the investor loses its principal or a portion of it. The definition of the 'occurrence', i.e. the trigger of the cat bond, can be stipulated in different ways: (a) by a certain loss to the sponsor (indemnity trigger); (b) by a market loss (industry loss index trigger); or (c) if a set of certain defined physical threshold values (e.g. wind speeds at certain points or discharges) are exceeded (parametric trigger).

Cat bonds are usually high-interest, but also high-risk investments, and are practically only purchased by professional investors. Investors choose to invest in catastrophe bonds because their return is largely uncorrelated with the return on other investments, and so cat bonds help investors achieve diversification. Investors who participate in this market include hedge funds, specialized disaster-oriented funds, asset managers, life insurers, reinsurers, banks and pension funds.

Cat bonds are geared to very large potential losses and are therefore an instrument exclusively used in developed countries. Another instrument used in a similar way is a swap. Here, an identical portion of risk from two independent (and distant) events is exchanged between two partners who are thus both sponsor and investor. For example, a portion of the 'Flood Europe' risk is transferred in exchange for the same portion of the 'Hurricane Florida' risk.

Conclusions

The Swiss writer Max Frisch once stated (Frisch, 1981:103): 'Only man knows natural disasters, so far as he survives them. Nature does not know disasters.' Such disasters have become more frequent and more intense during the last decades. The main causes are the increasing global population and its need – and sometimes wish – to settle in areas that are prone to natural hazards, often in highly concentrated urban agglomerations. At the same time, the amount and susceptibility of possessions grow as risk awareness fades.

The latest IPCC report (2013) has clearly documented that climate change is real, increases the risk in many regions and is caused by human activities. It is today one of the greatest risks facing societies, and the rising number of severe weather-related natural disasters will cause higher loss burdens for economies in the future.

The dramatic trend of disaster losses can only be lessened by a package of integrated countermeasures including strict land use regulations, warning systems, education, (financial) motivation and insurance. These measures have to be supported by all the people and institutions involved: public authorities, scientists and relief organizations, individuals and enterprises, insurance and reinsurance companies. Requirements must be formulated and prepared in such a way that the political powers can derive clearly recognizable policy options from them.

Society must understand that using tax revenue for the purposes of relief is nothing but subsidization – and of the more expensive kind, because losses cost more than precautions. At the same time, loss prevention measures must be enforced, the most important being the strict prohibition of building in high-risk areas. In this context, requiring insurance is a wise way to curb undesirable development. For no other peril are protection measures as effective with respect to risk and loss reduction as for floods. Numerous examples have proved that investment in protection and prevention pays off. It is also possible to establish an insurance system that is compatible with the needs of a given society if certain conditions are accepted and the various aspects are considered within a long-term perspective, and not on a day-to-day basis.

Great natural events are not avoidable. Great disasters are. Disasters are inevitably the net result of the effects of extreme natural events and the response to those events. Effective safeguards are both achievable and indispensable, but they will never provide complete protection. The determining factor is the awareness that nature can always come up with events against which no human means can prevail. If we are willing to tackle the risk from frequent events in a joint effort involving governments and the people, and if we are sufficiently prepared for the residual risk from rare events by involving the finance industry, we are headed for a world that becomes a safer place to live in.

References

Creel, L. (2003). *Ripple Effects: Population and Coastal Regions*. Population Reference Bureau Measure Communication, Washington, DC. http://www.prb.org/Publications/PolicyBriefs/RippleEffects PopulationandCoastalRegions.aspx.

Delli-Priscoli, J. & Stakhiv, E. (2015). Water-Related Disaster Risk Reduction (DRR) Management in the United States: Floods and Storm Surges. *Water Policy*, *17* (suppl.1), 58–88.

Demographia (2014). *Built-Up Urban Areas or World Agglomerations*, 10th edn. Demographia, Belleville, IL. Retrieved 25 July 2014. http://www.demographia.com/.

Frisch, M. (1981). *Der Mensch erscheint im Holozän*. Suhrkamp, Frankfurt.

IPCC (Intergovernmental Panel on Climate Change) (2013). Climate Change 2013: The Physical Science Basis. In: *Contribution of Working Group I to the Fifth Assessment Report of the Intergovernmental Panel on Climate Change*. Stocker, T. F., Qin, D., Plattner, G.-K., Tignor, M., Allen, S. K., Boschung, J., Nauels, A., Xia, Y., Bex, V., Midgley, P. M. (eds.). Cambridge University Press, Cambridge, 1535 pp.

Kron, W. (2005). Flood Risk = Hazard • values • vulnerability. *Water International 30*, 58–68.

Kron, W. (2012). Inland Floods. In: *Severe Weather in North America*. Kron, W. & Wöst, F. (eds). Munich Reinsurance Company, Munich, pp. 78–105.

Kron, W. (2013). Coasts – the high-risk areas of the world. *Natural Hazards 66* (3), 1363–1382. doi:10.1007/s11069-012-0215-4.

Kron, W. (2014). *Benefits Far Outweigh the Costs*. Topics Schadenspiegel 1/2014, Munich Re, Munich, pp. 40–44.

Kron, W., Steuer, M., Löw, P. & Wirtz, A. (2012). How to deal properly with a natural catastrophe database – analysis of flood losses. *Natural Hazards and Earth Systems Sciences (NHESS) 12*, 535–550. doi:10.5194/nhess-12–535–2012.

Lichter, M., Vafeidis, A., Nicholls, R. J. & Kaiser, G. (2010). Exploring data-related uncertainties in analyses of land area and population in the 'Low-Elevation Coastal Zone' (LECZ). *Journal of Coastal Research 27* (4), 757–768. doi:10.2112/JCOASTRES-D-10–00072.1.

Melecky, M. & Raddatz, C. (2011). *How Do Governments Respond After Catastrophes?* Natural-Disaster Shocks and the Fiscal Stance. Policy Research Working Paper 5564, World Bank, Washington, DC.

Munich Re (Munich Reinsurance Company) (2012a). *Topics Geo Natural Catastrophes 2011 – Analyses, Assessments, Positions*. Munich Re, Munich, 56 pp.

Munich Re (Munich Reinsurance Company) (2012b). *Severe Weather in North America*. Munich Re, Munich, 274 pp.

Munich Re (Munich Reinsurance Company) (2013). *Severe Weather in Eastern Asia*. Munich Re, Munich, 408 pp.

Munich Re (Munich Reinsurance Company) (2014). *Topics Geo Natural Catastrophes 2013 – Analyses, Assessments, Positions*. Munich Re, Munich, 60 pp.

Nicholls, R. J., Hanson, S., Herweijer, C., Patmore, N., Hallegatte, S., Corfee-Morlot, J., Chateau, J. & Muir-Wood, R. (2008). Ranking Port Cities with High Exposure and Vulnerability to Climate Extremes: Exposure Estimates. OECD Environment Working Papers, No. 1. OECD Publishing, Paris. 63 pp. DOI: http://dx.doi.org/10.1787/011766488208.

Small, C. & Nicholls, R. J. (2003). A global analysis of human settlement in coastal zones. *Journal of Coastal Research 19* (3), 584–599.

von Peter, G., von Dahlen, S. & Saxena, S. (2012). Unmitigated disasters? New evidence on the macroeconomic cost of natural catastrophes. *BIS Working Papers No. 394*. Bank for International Settlements, Basel, Switzerland.

Wieriks, K. & Schultz van Haegen, M. (2015). The Deltaplan Revisited – Changing Perspectives in the Netherland's Flood Risk Reduction Philosophy. *Water Policy*, *17* (suppl.1), 41–57.

Wirtz, A., Kron, W., Löw, P. & Steuer, M. (2012). The need for data: Natural disasters and the challenges of database management. *Natural Hazards 40*, 135–157. doi:10.1007/s11069-012-0312-4.

Lessons learned from mega-disasters and future policy development on water-related disaster management in Japan

Kenichiro Tachi

Director for International Coordination of River Engineering International Affairs Office, River Planning Division, Water and Disaster Management Bureau Ministry of Land, Infrastructure, Transport and Tourism (MLIT), Tokyo, Japan
E-mail: tachi-k2jy@mlit.go.jp

Abstract

Japanese disaster management approaches have been enhanced and strengthened through repeated experiences of disasters in the past. The report presents lessons learned from the Great East Japan Earthquake and Tsunami, a disaster caused by a low-probability but extremely large hazard, and introduces the distinct features of Japanese water-related disaster management, including those enhanced and strengthened based on the recent lessons. Finally, drawing from Japan's experiences, messages to be sent to the world from the water-related disaster community are proposed.

Keywords: Disaster prevention; Great East Japan Earthquake and Tsunami; Japan; Mega-disaster; Policy development; Structural measures; Water-related disaster management system

1. Introduction

Japan's topography and climate give rise to certain challenges. The steeply mountainous nature of most the country's territory leads to concentration of population and industry on limited, narrow alluvial plains, and the typhoons and weather fronts make torrential rain a frequent occurrence. Japan has in consequence experienced floods and numerous other water-related disasters. In these circumstances, it has been the continuous implementation of water-related disaster management measures corresponding to society's needs that has supported Japan's continued development and establishment of its social stability.

Water-related disaster management in Japan has developed through the implementation of measures based on the principle of preventing the recurrence of disasters and leaving a more disaster-resilient nation to posterity, learning lessons from repeated experiences of disasters. One distinctive feature of Japan's approach to managing water-related disasters is that it deploys an appropriate combination of

doi: 10.2166/wp.2015.002

hard (structural) and soft (non-structural) measures at all stages of disaster management (pre-disaster, emergency response, and post-disaster reconstruction and recovery). Another feature of the approach is that it does not largely rely on disaster response after the event; rather, it places emphasis on well-planned implementation of preventive measures at the pre-disaster stage. Both the legal framework and the technologies underpin water-related disaster management in Japan, which has undergone continual improvement as a result of numerous experiences of disaster.

During the Great East Japan Earthquake and Tsunami, Japan's territory was subjected to low-probability but extremely large hazards: the great seismic motion and the tsunami. While this disaster demonstrated that Japan's disaster management measures had been effective to a certain extent, it also brought us a lesson in that we had not adequately prepared for a situation in which the extent of the hazards greatly exceeded the design capacity of coastal levees and other structures. Learning from the lessons, understanding the importance of the 'assumption of hazards of any possible magnitude', new water-related disaster management approaches are being introduced in Japan. Thus, Japan is continuing on the same road that it has followed in the past, the road of learning from previous disasters and upgrading its disaster management capability.

The generating mechanisms of meteorologically or hydrologically induced water-related disasters are different from those of earthquakes and tsunamis. However, the issues and the countermeasures drawn from the Great East Japan Earthquake and Tsunami have enough generality and can be applied to other water-related disasters including hydrological and meteorological ones. Accordingly, in this study, the Great East Japan Earthquake and Tsunami and the lessons learned are described first. The distinctive features of Japan's approach to water-related disaster management, which has been enhanced through repeated experiences of disasters in the past, including those from the earthquake and tsunami, will be systematically presented together with details of the legal framework and the technologies that underpin the approaches. Our world is now facing the risks of both climate change and mega-earthquakes. Given this situation, we believe that, by sharing our experiences with other countries, Japan can make a major contribution to improving the world's disaster management capability.

2. The Great East Japan Earthquake and Tsunami and the lessons learned

2.1. A brief account of the earthquake

On 11 March 2011, a 9.0 magnitude earthquake occurred in the Pacific Ocean off the coast of Tohoku, causing major seismic motion across a large area of Japan. Approximately 30 minutes after the earthquake, a giant tsunami struck a 650-km stretch of the Tohoku coastline, inundating an area of over 500 km^2.

The tsunami left about 20,000 people dead or missing, and caused the total collapse of about 130,000 buildings, the partial collapse of about another 260,000 buildings, and damage to about 730,000 more buildings. The damage from the disaster was immense, with up to about 470,000 people evacuated, while the economic losses amounted to 16.9 trillion yen.

The distinctive features of the Great East Japan Earthquake and Tsunami were the low probability but massive scale of the earthquake and the coincidence of the tsunami hazard with the seismic vibration. The earthquake and tsunami resulted in damage to an extremely large area and caused extensive knock-on effects including power outages, nuclear accidents, and repercussions for industry worldwide through the supply chain. For details of the Great East Japan Earthquake and Tsunami and lessons learned, refer to Ranghieri & Ishiwatari (2014).

2.2. *Lessons from the Great East Japan Earthquake and Tsunami*

The lessons learned from the Great East Japan Earthquake and Tsunami are described below.

1. Measures developed after the Great Hanshin-Awaji Earthquake (1995), including the earthquake-proofing of buildings and infrastructure, were effective against large seismic motion to a certain extent.

In January 1995, Japan was hit by the Great Hanshin-Awaji Earthquake, which left about 6,500 people dead or missing. The majority of the victims of this earthquake died when buildings constructed according to the old (pre-1981) earthquake-resistance standards collapsed, followed by a lesser number of those who died in the fires that broke out after the collapse. In addition, sections of the elevated Kobe expressway, the region's arterial road, buckled or collapsed. In response to the lessons learned from this disaster, Japan has carried out a systematic program to promote earthquake-proofing in buildings and seismic reinforcement of the country's elevated roads. Consequently, the buildings that collapsed during the Great East Japan Earthquake and Tsunami were mainly structures built according to the old earthquake-resistance standards. Despite the extremely large seismic motion, the majority of buildings having undergone adequate seismic reinforcement escaped damage. Also, bridges that had been seismically reinforced did not collapse or sustain other serious damage, which made it possible for these to be used as escape and relief routes as the authorities were able to bring them back into service soon after the earthquake (see Figure 1).

A distinctive feature of disaster management in Japan, the systematic implementation of preventive measures in the pre-disaster phase, can be considered a success. Continuous investment in such measures can consequently be regarded as essential.

2. Preparedness for a tsunami that greatly exceeded the design capacity of the defense structures was not sufficient. It is important to mitigate human and economic losses by assuming possible hazards of maximum scales and deploying 'multi-layered protection', combining structural and non-structural measures.

Fig. 1. Effect of seismic reinforcement of bridges after the Great Hanshin-Awaji Earthquake.

Since the height of the tsunami that struck the coastal area during the Great East Japan Earthquake and Tsunami greatly exceeded the design height of the defense structures, the tsunami overtopped coastal levees, and inundated a large area inland. Although levees and other facilities in some areas did retard the advancing tsunami and reduce the extent and depth of inundation, many of them collapsed after they were overtopped. Advancing at high speed, the waters of the tsunami inundated inland areas to a great depth within a short time, and swept away numerous buildings and other property, causing extensive damage and claiming many lives.

As we reflect on these events, it is recognized all the more clearly that 'hazards can be of any possible size', thereby highlighting the importance of implementing all possible measures to 'protect lives whatever it takes' even if a tsunami of the maximum-possible scale occurs. Accordingly, policy was developed for tsunami protection measures, applicable also to reconstruction and recovery work after the Great East Japan Earthquake and Tsunami (see Figure 2). As referred to in Section 3.4.2.1 (ii), the Act on Regional Development for Tsunami Disaster Prevention became law in December 2011 to actually implement this policy. The policy aims to promote tsunami countermeasures of 'multi-layered protection', which is to flexibly apply structural and non-structural measures in comprehensive regional development. Against comparatively frequent tsunamis with certain scales, structures such as coastal levees will be developed to protect lives and assets, and to conserve national land. Against tsunamis with maximum scales that would cause devastating damage, although the frequency of occurrence is quite low, non-structural measures will be applied in addition to the structural measures for mitigation of damage. Taking into account the characteristics of each region, existing public and private facilities are fully utilized in the application of measures. Under the said concept, the Act on Regional Development for Tsunami Disaster Prevention became law in December 2011. In the development of coastal levees, to cope with tsunamis of a greater scale than the defense-design height, technological development will be promoted to make structures resiliently continue to exert their effects even after being overtopped. This example of the new policy development illustrates a distinctive feature of disaster management in Japan: the strategy of continuous improvement of both a legal framework (Tsunami Disaster Management and Regional Development Act) and technologies (technological development of resilient coastal levee structures) in response to the lessons learned from previous disasters.

3. An extensive transportation network provided alternative routes for rescue activities and transportation of relief goods, and road embankments were effective in preventing tsunami propagation further inland. It is important to incorporate or mainstream disaster risk reduction in every social system.

In the extensively damaged coastal areas, many roads including arterial routes were cut off by piles of debris or structural damage. Keeping routes to the coast open for rescue and relief operations was a matter of life and death. Accordingly, in the aftermath of the disaster, the Tohoku Regional Bureau of the Ministry of Land, Infrastructure, Transport and Tourism immediately began clearing roads that run from longitudinal roads, which pass through inland areas where damage was less severe, to the coastal regions. Eleven routes had been cleared within one day after the disaster and fifteen routes within four days after it, providing access for ambulances and police, the Self-Defense Force and other emergency-service vehicles. For details of this operation, refer to Tohoku Regional Bureau, Ministry of Land, Infrastructure, Transport and Tourism (2014).

The Sanriku Expressway, which runs along the coast of Tohoku region, was planned to run through a hilly area in view of the effects of the tsunami. As a result, it functioned as an evacuation area for neighboring residents and as a road for emergency transport as an alternative to the damaged Route 45, which

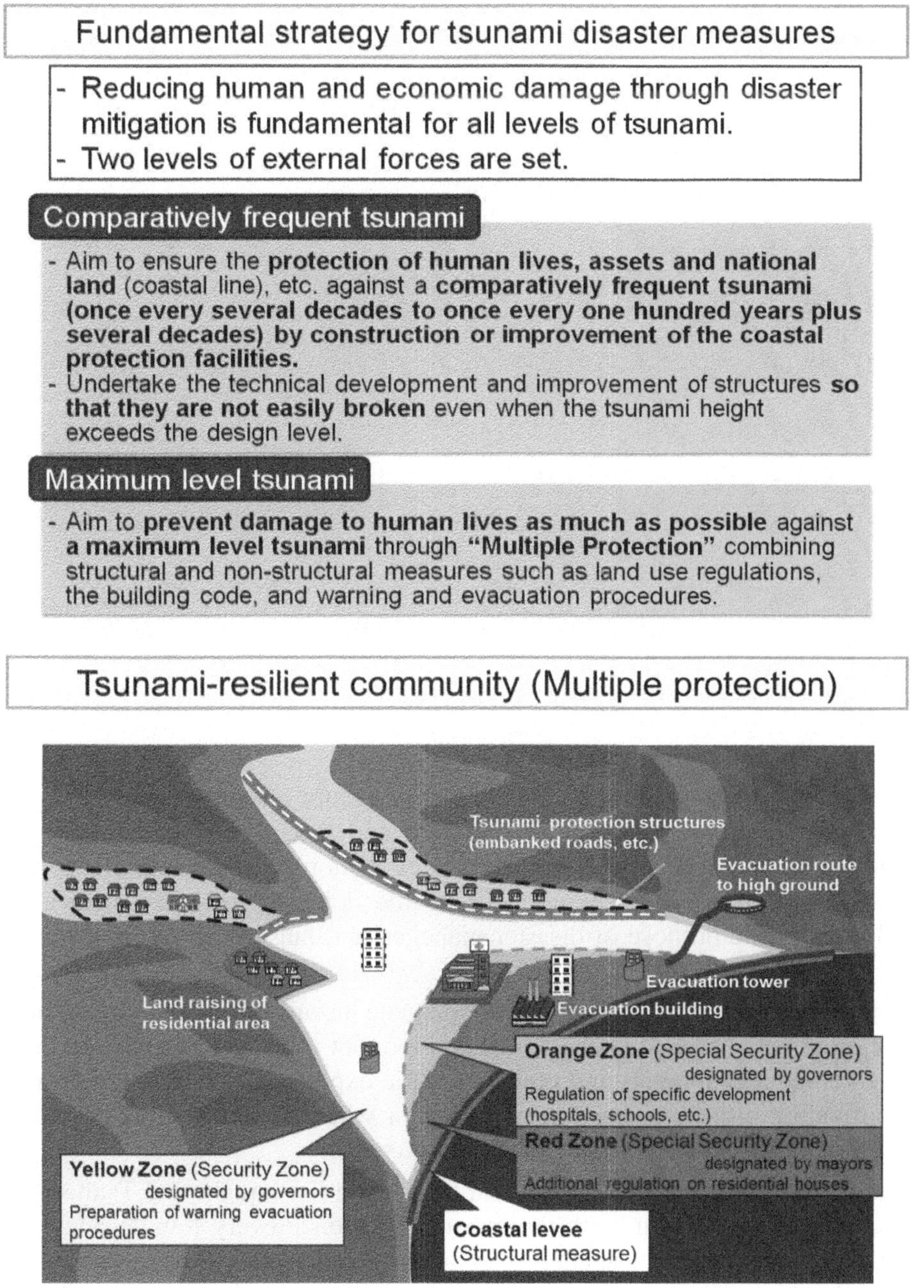

Fig. 2. New policy development for tsunami prevention measures.

runs along the coast. The East Sendai Expressway, built on an embankment 7–10 m above its surroundings, obstructed the flow of the tsunami, which had traveled almost 4 km inland across the Sendai Plain, thus functioning as a barrier to the advancing tsunami water and debris. Furthermore, local residents were able to escape onto the embankment (see Figure 3).

Fig. 3. An expressway functioned as a tsunami barrier and an evacuation site.

These examples show that expressways developed to secure alternative routes support rescue and relief operations and recovery activities after disasters, and that road embankments work as barriers to limit the inundation area. The aforementioned cases demonstrate the importance of incorporating the perspective of disaster management not only in the defense structures, such as coastal levees, which are developed for the purpose of disaster management, but also in all of society's systems.

4. Records of past disasters and the provision of adequate hazard information to the public in advance supported appropriate evacuation behavior. It is important to collect and pass on disaster records to others, and to apply technologies and other scientific expertise, including those of inundation prediction, in providing disaster protection structures or when pursuing urban development.

The region damaged by the tsunami in the Great East Japan Earthquake and Tsunami had been recurrently hit by the same phenomenon, with the frequency of once every few decades. This means that knowledge of the threat posed by tsunamis had been passed down from generation to generation, and that local people were well aware of the significance of preparing for disasters. There are stone markers scattered all over the region, commemorating the damage incurred by past tsunamis or showing the height that they reached. Without doubt, these played a part in keeping residents aware of the importance of preparing for the tsunamis that occur once every few decades.

In Kamaishi City, where the tsunami claimed about 1,000 victims out of a population of about 40,000, elementary and junior high-school students had been taking part in repeated evacuation drills and disaster-education activities using tsunami hazard maps. As a result, the death toll for elementary and junior high-school children was comparatively low, with five fatalities out of approximately 2,900 children in these age groups. The death rate was about five percent of that of the whole city.

Conversely, the occurrence of the gigantic tsunami and the consequent extensive damage revealed the limit of disaster countermeasures excessively dependent on structure development, even though the structures such as coastal levees are effective against tsunamis that do not exceed the defense-design height. It cannot be denied that the existence of such structures gave people a false sense of security due to the inadequate communication about the risks. Also, the tsunami hazard maps made no reference to a low-frequency mega-tsunami, which occurred on this occasion, and in many places, the actual inundation area was far greater than the area shown on the map. As a result, evacuation behavior may not have been based on an accurate understanding of the local risks arising from the disaster.

There are limitations in *ex-ante* risk assessment technologies. It is important to develop techniques to enable assessment that is based on more diverse multiple scenarios, by collecting a wider range of data, including the analyses of historical documents, and so on. Moreover, the success of the evacuation in Kamaishi City proved awareness-raising, education and evacuation drills used in normal times to be effective. Knowledge-sharing on initiatives that enable effective use of hazard maps is as important as that of producing the hazard maps. Adequately assessing the risks and accurately informing local residents about these risks are indispensable for appropriate evacuation behavior and the development of a community's disaster-resilience. When 'multi-layered protection' based on the 'assumption that the hazards can be of any possible size' is being developed for future tsunami protection, initiatives for the aforementioned purposes must be stepped up. In tandem with improvements in the accuracy of flood prediction and other technologies, there must be more consistent application of the technological knowledge to assess the risks and provide disaster-risk management in an effective manner. For this purpose, it is important to make and pass on to others, not only the accurate records of the inundated area and its depth, and so on, but also detailed records of the activities of residents and relevant administrative bodies.

Passing on knowledge of relatively low-frequency disasters such as the tsunami is not an easy task. However, various initiatives are under way in the affected areas to collect, store and pass on records of disasters.

As stated above, many lessons have been learned from the Great East Japan Earthquake and Tsunami. Recovery and reconstruction are being progressed, with a view to develop safer regions for posterity, making use of the lessons learned from the disaster.

Japan has only just begun the process of improving its disaster management in response to the lessons learned from the Great East Japan Earthquake and Tsunami. The evaluation of the results of this process will be a task for posterity. In any case, Japan should continue on the road of continuously improving its legal framework and technologies in the light of past experiences of disaster.

3. Japan's water-related disaster management system: learning lessons from numerous and diverse disasters

3.1. Overview of Japan's water-related disaster management system

The previous section reviewed the lessons learned from the Great East Japan Earthquake and Tsunami and described how disaster management in Japan is being enhanced in response to these lessons. As stated in that section, disaster management in Japan has been strengthened in the light of repeated experience of earthquakes, mega-water disasters and other events. Indeed, it can be said that disaster management in Japan has been shaped by a process of continuous improvement. Here, we outline the special features of disaster management in Japan, in particular water-related disaster management, and how this has been shaped by the country's extensive experience of disasters.

Since one distinctive feature of Japan's water-related disaster management system is that it deploys an appropriate combination of structural and non-structural measures at all stages of disaster management (pre-disaster preparedness, emergency response and post-disaster recovery and reconstruction), we will describe here measures for each phase in chronological order. Also, since disaster management in Japan has, in the light of repeated experiences of disaster, been supported by continuous improvement of its legal framework and technologies, the major aspects of this framework and the most important of these technologies are described below.

3.2. Pre-disaster

3.2.1. Well-planned investment in disaster management measures in the pre-disaster phase. On the basis of assessment of potential damage, disaster management plans are made and measures are consistently implemented in the pre-disaster phase. These measures form a multiple defense, appropriately combining structural and non-structural components. The measures include construction of embankments and other river improvements; preparation and publication of hazard maps to provide guidance on safe land use and to facilitate flood response measures by residents, businesses and other entities; construction of systems for early warning based on flood forecasts and other information; evacuation and disaster response training and drills; and so forth.

3.2.1.1. Legal framework

(i) The River Law provides for the establishment of a flood-control system (under long-term 'fundamental river management policies' and 'river improvement plans'). The purpose of the River Law is 'to contribute to land conservation and development, and thereby maintain public safety and promote public welfare by administering rivers comprehensively to prevent occurrence of damage due to floods, tsunamis, storm surges, etc., ensure appropriate river use, maintain the normal functions of the river water, and improve and conserve the fluvial environment'. The river administrator designated under this law (for example, the Ministry of Land, Infrastructure, Transport and Tourism in the case of class A rivers) is authorized to draw up long-term 'fundamental river management policies' and short- and medium-term 'river improvement plans', and to carry out river improvement based on these plans. Generally, river improvement plans are drawn up after consideration of the views of academic experts, local residents and local mayors. The fact that these plans are both based on the law and drafted with consideration for local opinion ensures proper implementation of the measures that these plans provide for.

(ii) The Flood Fighting Law provides for the support of initiatives by local communities against floods. The River Law mainly provides for long-term structural measures. In contrast, the Flood Fighting Law provides for non-structural measures including actions by flood fighting organizations; transmission of flood forecasting information to local residents by administrative agencies; provision of risk information through the designation of areas prone to flooding and publication of flood hazard maps; promotion of the participation of residents' and businesses' flood fighting activities;

and implementation of other necessary measures. The River Law and the Flood Fighting Law operate in tandem to promote integrated structural and non-structural measures.

3.2.1.2. Disaster management technology

(i) Run-off and inundation prediction technologies are used for analyzing cost-effectiveness of investments at the pre-disaster stage, risk assessment, and so on. Whenever plans for river improvement are being prepared, water levels and river discharges for floods are calculated using run-off and hydraulic analysis based on observed values of accumulated precipitation and river discharge. In the assessment process for flood control projects, areas susceptible to inundation are identified by flood inundation analyses, and the benefits resulting from flood control measures are evaluated. In addition, flood inundation analysis technologies are used to identify the possible impacts of inundation (inundation damage risk assessment) and so forth. In recent years, in order to adequately comprehend flood disaster risks, quantitative assessment of damage categories that were not conventionally assessed due to difficulties in monetization has been introduced in the assessment process for flood control projects. Human damage, impairment of social function, and cascading economic impacts are among the newly introduced damage categories (see Figure 4).

Use of the above-named analytical technologies facilitates efficient and effective project planning and implementation through quantitative assessment of the effectiveness of flood control measures.

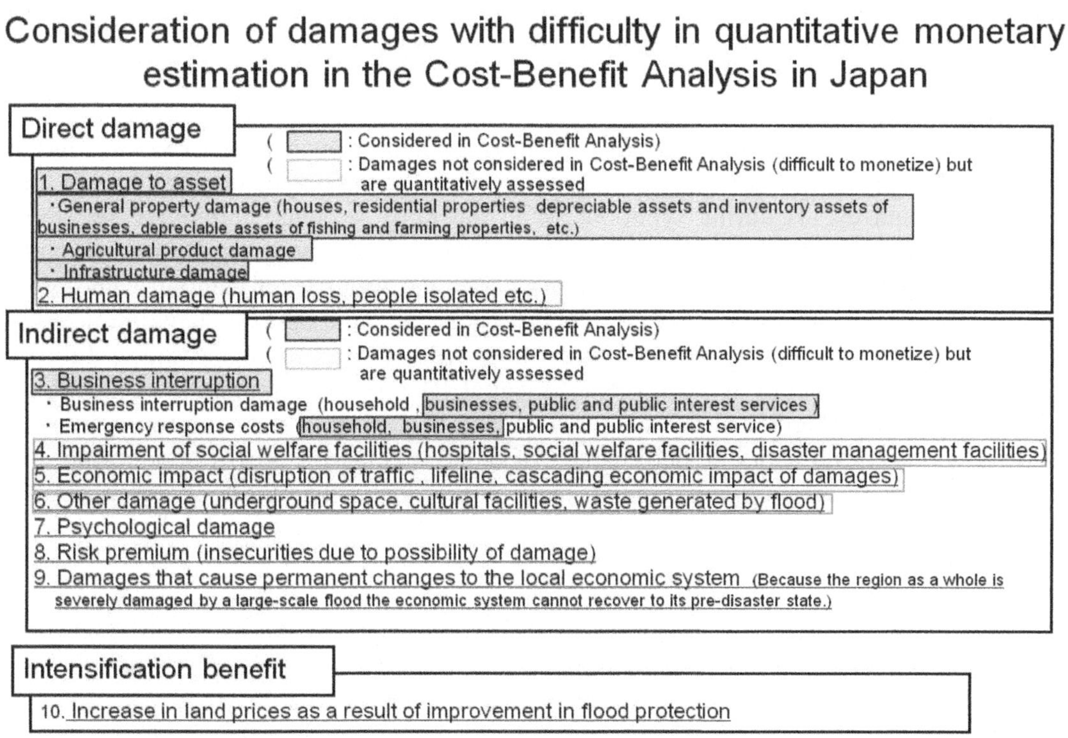

Fig. 4. Newly introduced damage categories in the assessment of flood control projects.

3.3. Emergency response

3.3.1. Appropriate infrastructure operation and other emergency responses. Emergency responses include the appropriate operation of dams, sluice gates and other types of infrastructure based on rainfall and water level information; the provision of flood and other relevant information to assist flood fighting activities and the evacuation of residents; rapid removal of flood water using drain pump vehicles; and other measures as appropriate.

3.3.1.1. Legal framework

(i) The river management framework secures responsible local offices for river maintenance and management. The existence of designated 'public facility administrators' is an important feature of Japan's public infrastructure management system. In the case of a river, the River Law designates a 'river administrator', such as the Minister of Land, Infrastructure, Transport and Tourism for class A rivers. In normal circumstances, the river administrator is responsible for river-related studies and planning; river patrols; maintenance and management of dams, sluice gates and other types of infrastructure; construction work on embankments and other defense structures; and conservation and improvement of the fluvial environment. During flood events, the administrator is responsible for tasks including monitoring precipitation and water levels; providing flood forecasts and other relevant information; operating dams, sluice gates and other types of infrastructure; ensuring rapid removal of flood water using drain pump vehicles and other methods; and post-disaster surveys. Making a single entity responsible for infrastructure management during both normal circumstances and flood events fosters the accumulation of knowledge and experience relating to rivers and river basins, and the development of a sense of responsibility, thus facilitating appropriate responses to an emergency.

3.3.1.2. Disaster management technology

(i) Forecasting and warning technology enables infrastructure management based on hydrological forecasting, assisted by real-time data transmission systems and radar technology. Japan frequently experiences heavy rain events during typhoons or in the rainy seasons, and riverbank steep slopes cause quick flood run-off. For these reasons, it is important to accurately monitor such hydrological and hydraulic information as precipitation, river discharge and water level without delay, provide relevant information to the authorities and residents concerned, and operate dams, sluice gates and other types of infrastructure appropriately using the information. Also, to enable successful evacuation by residents, accurately predicting water levels is essential. The response to rapidly occurring emergencies on Japan's rivers is facilitated by the deployment of real-time data transmission systems of precipitation and water level data, the deployment of radar systems for precise monitoring of precipitation distribution, and forecasting and warning technologies that utilize these observed data.

3.4. Post-disaster response

3.4.1. Surveying, analyzing and recording damage. Surveys, analyses and records of damage are used for rapid planning of recovery and reconstruction. After a flood event, river administrators must survey and collect such information as flood high-water marks (maximum river water levels during a flood), flood inundation area, and incurred damage. The results of surveys are used for improving the analytical methods that are essential in river improvement planning and so forth. They are also reflected in the improvement of plans for river improvement works and maintenance. Data regarding flood damage are indispensable for promoting public understanding of the need for flood control measures.

3.4.1.1. Legal framework

(i) The water disaster statistical survey system is an important data-compiling system. Under the Statistics Act, since 1961, surveys of damage to personal or business property, rivers, roads, or other public infrastructure, transport, telecommunications, or other public services from floods, inundation by insufficient drainage, storm surges, or debris flows have been carried out with the aim of compiling basic data required for the implementation of various flood control policies. It has become an institutionalized procedure that the municipal governments conduct surveys and reports on damage to personal and business property, and the prefectural governments conduct surveys and reports on damage to public infrastructure (see Figure 5). These surveys and reports help to build up reliable and consistent records of water-related disaster damage.

3.4.2. Recovery and reconstruction for future damage reduction. Preventing recurrence of disaster is a priority when rebuilding damaged infrastructure. Reconstruction in disaster-stricken regions provides these areas with greater reduction of disaster risks.

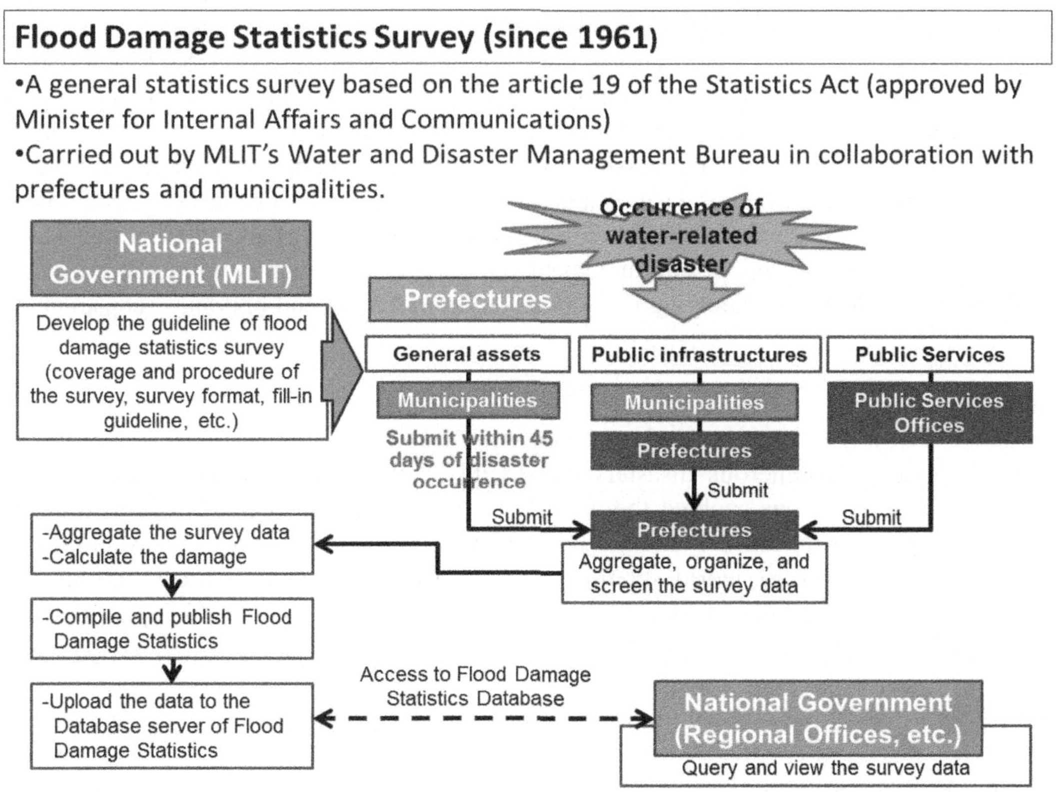

Fig. 5. Flood damage statistics survey since 1961.

3.4.2.1. Legal framework

(i) The first significant component of the legal framework is the post-disaster recovery projects system. Prompt and sound recovery of public infrastructure damaged in natural disasters is important for securing social welfare. However, the cost of rebuilding public infrastructure can burden local public bodies with impulsive and heavy spending. To address this issue, Japan has set up a system for recovery projects aimed at various kinds of public infrastructure, under which central government increases its share of these expenditures depending on the local public body's financial situation. Unlike as for other public works projects, funds are made available rapidly, in accordance with the project's urgent needs for recovery, and damage assessments for the purpose of determining the projects are carried out as soon as local public bodies are ready. The local public body can start reconstruction work immediately after a disaster before completion of the damage assessment. Furthermore, if restoring infrastructure to its pre-disaster condition is inappropriate or significantly difficult, then the reconstruction may include qualitative improvements in position, configuration, materials, dimension, and so forth. For example, a winding road that was washed away due to a sediment disaster may be replaced by another road built on a new route, such as a tunnel connecting both ends of the damaged section. Under this system, post-disaster recovery in Japan can take into account prevention of disaster recurrence.

(ii) The other significant component of the legal framework is the Act on Regional Development for Tsunami Disaster Prevention. As recovery continues in areas that suffered major damage during the Great East Japan Earthquake and Tsunami, regional development must promote resilience to tsunami disasters that may occur in the future. There was a need to establish a general system, covering the whole of Japan, that would be capable of preventing or mitigating any future tsunami disasters. In response to this need, the Act on Regional Development for Tsunami Disaster Prevention became law in December 2011. In accordance with this act, 'regional development for tsunami disaster prevention' is being promoted through a principle of 'multi-layered protection' combining structural and non-structural measures. The act authorizes prefectural governors to determine and announce the possible extent of tsunami inundation (including the areas thought to be at risk of flooding in the event of a tsunami and the estimated depth to which they would be flooded). Using these projections, municipalities can develop regional development plans in such a way as to promote integrated tsunami disaster prevention, and implement various measures such as setting up warning and evacuation systems for residents, and applying restrictions on development and buildings' construction.

4. Conclusion

Japan has experienced numerous disasters due to its challenging topography, and to its climate. Experiencing such disasters has enabled the country to improve its disaster management capability.

As the disaster risks from climate change and mega-earthquakes increase, the sharing of Japan's experiences of disaster with the world should contribute to enhancing the world's ability to address and manage disasters. In the light of its experiences, it would be appropriate for Japan to send the following message to the world.

(i) Strengthening disaster management investment for 'disaster prevention'

At present, most countries can hardly be said to be making sufficient investment in disaster prevention. In most countries, disaster management has remained at the stage of emergency response and relief during disaster events. However, disasters do not only claim many human lives. By destroying buildings and other property, they also sweep away the economic development gains so far achieved. In the developing countries, eradicating poverty is a priority task. It is necessary to understand that disaster prevention is important for laying the foundations of growth and protecting the fruits of development. Japan's experience of achieving economic development under challenging topographical and climatic conditions demonstrates that disaster prevention measures make economic sense, given the costs resulting from disaster damage.

(ii) Appropriately combining structural and non-structural measures

Early warning systems, disaster education, and other non-structural measures are effective in protecting human life when disasters occur. However, the evacuation action, the primary non-structural measure, cannot by itself protect property from disasters. This means that the development gains may be lost in disasters, which would result in delaying the achievement of the international goal of eradicating poverty.

Structural measures are effective not only against disasters that are within the design capacity of defense structures, but also, to a certain extent, against disasters exceeding this design capacity (see Figure 6). Furthermore, even when a disaster exceeds the design capacity of defense structures, an appropriate combination of structural and non-structural measures can minimize damage. In response to the lessons

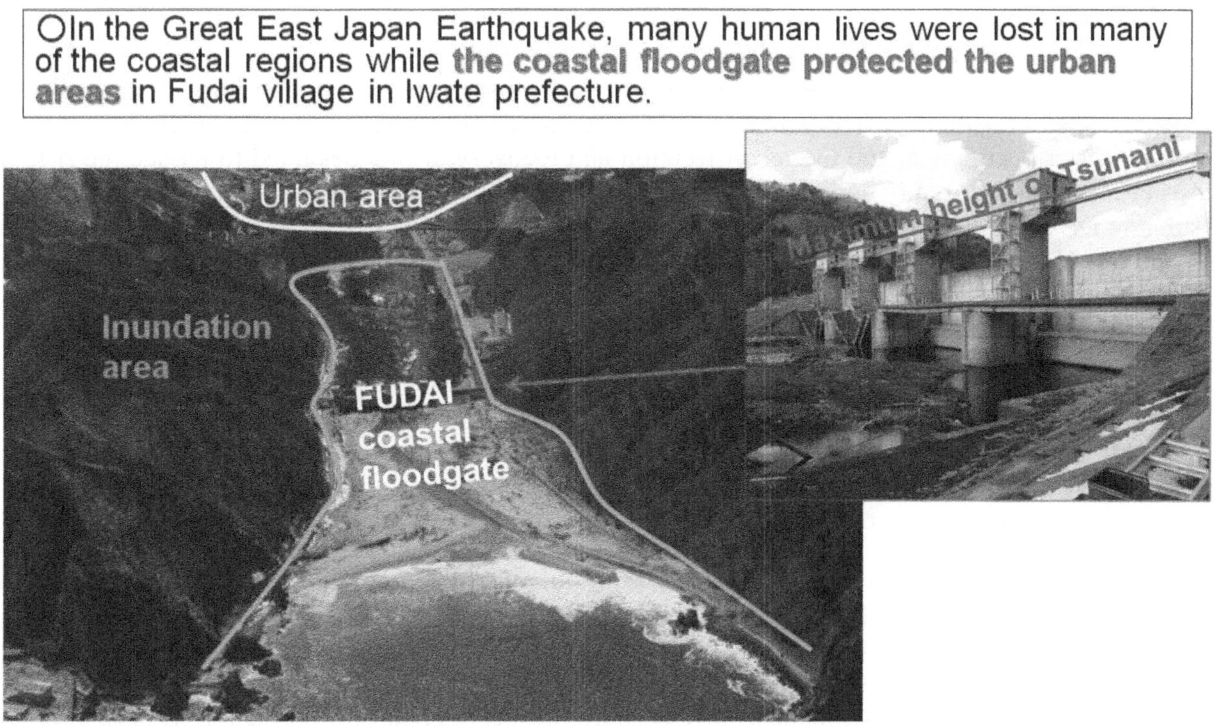

Fig. 6. An example of structural measures that provided protection against a tsunami that exceeded the defense-design capacity.

learned from the Great East Japan Earthquake and Tsunami, Japan is implementing both structural and non-structural measures as it pursues regional development for tsunami disaster prevention. Structural measures may be more expensive than non-structural measures, but, with the prevention of damage from future disasters in mind, an appropriate combination of both kinds of measure seems to be a wise investment.

(iii) Incorporating disaster management in all sectors

Building a disaster-resilient society cannot be achieved by projects with a 'disaster management' label alone. Disaster management needs to be incorporated into projects within other sectors also. During the Great East Japan Earthquake and Tsunami, the redundancy built into the road network, or the function of the road network to enable provision of alternative routes when there is disruption on some roads, clearly played a decisive role in rescue and relief efforts in the disaster aftermath. Moreover, road embankments acted as secondary levees. Although the focus in road development is on increasing the convenience of travel and providing an economic stimulus, we should be fully aware of the role that roads play when a disaster occurs. Having evaluated this role, we should seek to maximize all the benefits of road development including its disaster mitigation function. In many countries, budget limitations make it difficult to accord higher priority to investment in disaster management infrastructure. However, consideration of disaster management issues when investing in projects of any sector will enable efficient and effective enhancement of the disaster response capability of these countries.

(iv) Reconstruction and recovery planning for damage reduction

Japan's examples present good cases of transforming a disaster into an opportunity to leave a more disaster-resilient region to posterity, that is, the budget system allowing recovery to begin immediately after a disaster occurrence, and the reconstruction and recovery efforts designed to reduce the risk of disaster recurrence currently under way in the region affected by the Great East Japan Earthquake and Tsunami. Disasters catch the world's attention, and thus enable affected regions to receive a great deal of support. However, merely rebuilding the damaged infrastructure without implementing adequate measures during reconstruction and recovery efforts cannot reduce possible loss of life and property if another disaster happens. Disaster should not be treated as a transitory experience. In the long term, repeated efforts to prevent such an experience from recurring will build a disaster-resilient nation (see Figure 7).

(v) Assessing and recording damage

In many countries, there is still no system in place for continuous and consistent collection of disaster-related records and statistics. In Japan, there is a long-established system for collecting statistics on disasters including water-related disasters (see Figure 8). This is useful not only for assessing the effectiveness of disaster management measures and setting priorities; it also fosters understanding of the necessity for such measures and facilitates the securing of budgets and other resources, allowing these measures to be fully implemented. It is essential for every country in the world to have systems for continuous recording of disasters and collection of relevant statistics, so that the justification for

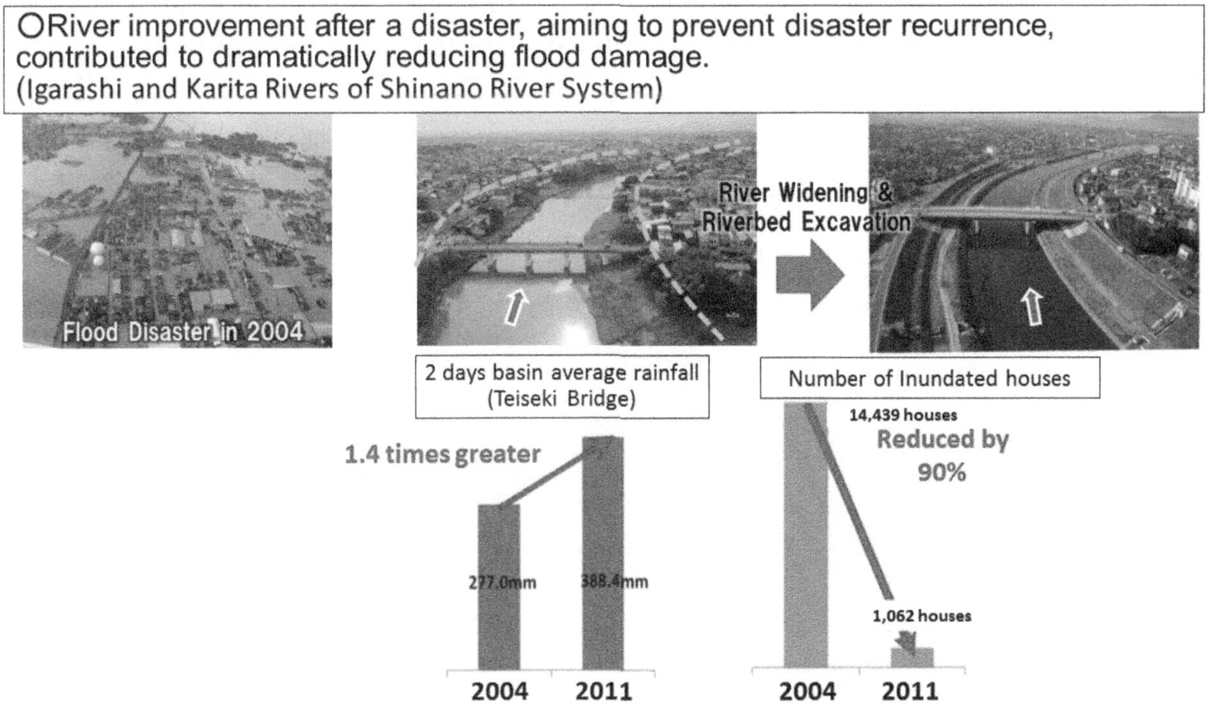

Fig. 7. An example of recovery and reconstruction works that prevented the recurrence of disaster.

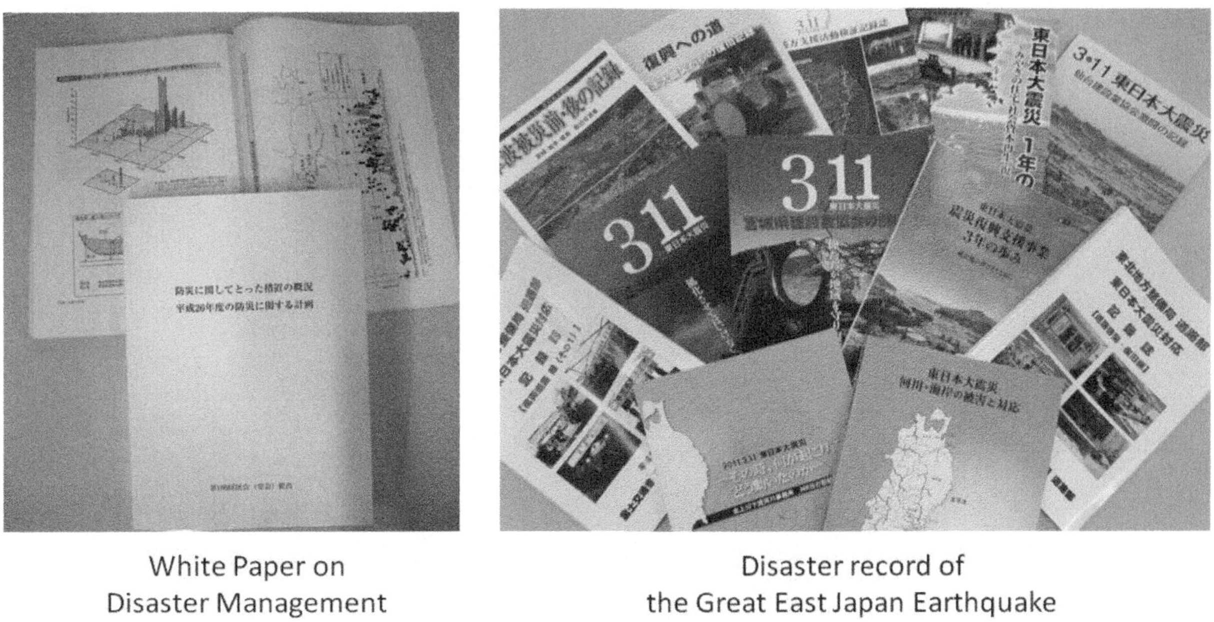

Fig. 8. Efforts to collect disaster-related records and statistics.

investing in disaster management can be adequately explained, and systems to enable implementation of the necessary measures created. It is vital that such initiatives are promoted worldwide.

References

Ranghieri, F. & Ishiwatari, M. (2014). *Learning from Megadisasters: Lessons from the Great East Japan Earthquake*. World Bank, Washington, DC.

Tohoku Regional Bureau, Ministry of Land, Infrastructure, Transport and Tourism (2014). *Leading the First Response to Large-scale Natural Disasters*.

The Deltaplan revisited: changing perspectives in the Netherlands' flood risk reduction philosophy

Melanie Schultz van Haegen[a] and Koos Wieriks[b]

[a]*Minister of Infrastructure and the Environment, The Netherlands*
[b]*Corresponding author. 1. UN Secretary General's Advisory Board for Water and Sanitation 2. Strategic Adviser, Ministry of Infrastructure and the Environment, The Netherlands*
E-mail: koos.wieriks@minienm.nl

Abstract

This paper focuses on the recently changed perspectives in the Netherlands' risk reduction philosophy, aimed at protecting a densely populated delta at a lasting high safety level throughout this century while addressing the impacts of climate change and socio-economic developments. The changed perspectives reflect a shift from policy development and implementation following a disaster (reactive) to a nationwide approach of climate-proofing the delta and thus anticipating projections of climate change and socio-economic developments at the end of this century (proactive: prevention). In addition to presenting these changed perspectives, lessons learnt are presented from Dutch experience, which can be applied to flood-prone areas around the globe. The connecting theme through these lessons learnt is the realisation that a proactive approach to flood prevention instead of reacting to flood disasters pays off and should be the way forward for densely populated, flood-prone areas.

Keywords: Adaptation; Deltaplan; Dutch approach; Flood risk management; Lessons learnt; Multi-level governance; Risk reduction; The Netherlands

1. Introduction

1.1. Worldwide increase of flood frequency and damage

Worldwide, the risk of weather-related disasters, such as floods and landslides, has increased strongly over recent decades (Figure 1; see also IPCC, 2014). Of all natural disasters most damage and most casualties are due to floods, especially river floods, which occur on a regular basis and cause ever more damage and affect many people. Not just the damage due to floods, but also the frequency of floods has increased, both in Europe and worldwide.

In its broad sense, flood risk is defined as the probability of adverse effects due to inundation. This means that risks will increase due to either an increase in frequency of extreme events, or because more

doi: 10.2166/wp.2015.003

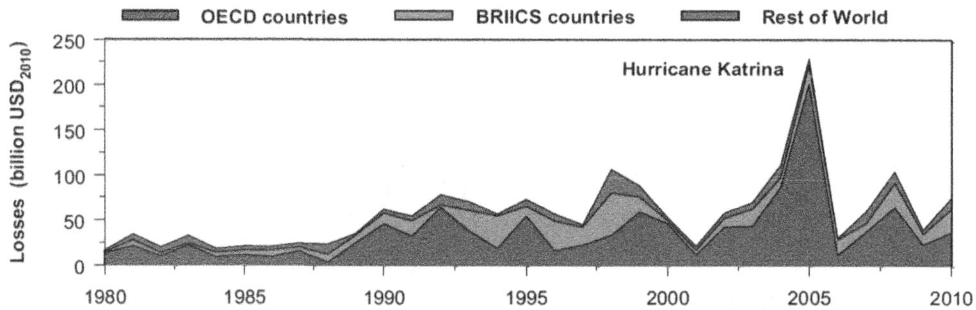

Fig. 1. The evolution of economic losses due to weather-related disasters over the last 30 years, stacked to three economic regions (Organisation for Economic Co-operation and Development [OECD] countries; Brazil, Russia, India, Indonesia, China, South Africa [BRIICS] countries; and Rest of World) (from Visser *et al.*, 2014)[1].

people or more economic assets are affected. Flood risk has already increased due to socio-economic developments over recent decades and will continue to do so due to further socio-economic developments and projected climate change. The risk increase over recent years has mainly resulted from the increase of the vulnerability of people and their belongings in flood-prone areas: more people have moved to low-lying areas near rivers and in the coastal zone (IPCC, 2014; Visser *et al.*, 2014).

The low-lying areas along the world's coasts constitute 'only' 2% of the world's land area but are inhabited by 10% of the world's population (and 15% of the world's urban population). Furthermore, 65% of the world's largest cities, with a population of at least 5 million residents, are located in the coastal zone. The global population exposed to the '1 in 100 year' extreme sea level (i.e. the sea level that has a 1% chance of being exceeded every year) has increased by 95% from 1970 to 2010 with about 270 million people and 13 trillion US$ worth of assets being exposed to this sea level in 2010. Compared to other regions, Asia exhibits the greatest exposure in terms of population and assets (IPCC, 2014).

1.2. High vulnerability, also in Europe

Also in Europe (river) floods occur on a regular basis. Over recent years large-scale floods have hit, among others, the Czech Republic and Germany (2002), Romania and Bulgaria (2005), the Alps (2005), England (2007), France (2010), Austria, the Czech Republic and Germany (2013), and recently Serbia and Bosnia Herzegovina (2014). The reported economic losses due to these floods vary from €1.7 billion (Romania, 2005) to €20 billion (the Czech Republic and Germany, 2002). Overall, there were dozens of fatalities per flood, up to over 100 in 2002 (Ulbrich *et al.*, 2003; Habersack & Krapesch, 2006; DETEC, 2008; European Environment Agency (EEA), JRC & WHO, 2008; Pitt Review Team, 2008; Pollner *et al.*, 2008; Săvoiu, 2008; Cour des Comptes, 2012; Conradt *et al.*, 2013).

These numbers and the frequency of occurrence of these large-scale floods illustrate that even in (Western) Europe the vulnerability to flooding is high in most countries. These numbers also demonstrate that flood protection pays off: the costs of the measures that have been taken in the Netherlands over recent years to increase the discharge capacity of the major rivers (the so-called Room for the River Programme), a total of €2.3 billion, seem relatively modest when compared to the recent flood damages of up to €20 billion. These measures reduce flood risk for some 4 million people.

---

[1] In this paper, the term 'billion' corresponds to 10^9.

In the scientific literature on risk reduction the terms 'vulnerability' and 'resilience' are often used. Vulnerability may be defined as '*the characteristics of a group and their situation that influence their capacity to anticipate, cope with, resist and recover from the impact of a natural hazard*' (Wisner *et al.*, 2005). According to the scientific literature the vulnerability to a hazard can be reduced by increasing society's resilience. A society is more resilient when it recovers more easily from a disaster, learns from the consequences ('build back better') and benefits from these 'lessons learnt' to anticipate future threats (Wisner *et al.*, 2005; IPCC, 2012, 2014). Improving your resilience is therefore a sustainable strategy to decrease your vulnerability.

1.3. Changed perspectives in the Netherlands

Flood protection in the Netherlands used to be based on a flood defence system of dikes, dams and storm-surge barriers that are designed and maintained at high safety standards put down in law. It has become clear though that these safety standards no longer reflect the value of the property and the number of people behind the dikes (Ten Brinke *et al.*, 2008). It was also concluded that people living in the flood-prone areas of the Netherlands are very vulnerable to flooding: evacuating people out of the flood-prone areas in the case of a dike failure is not an option for most of these areas (Kolen, 2013), and the number of casualties during one flood may be very high (Planbureau voor de Leeform geving/Netherlands Environmental Assessment Agency (PBL, 2014). From this understanding of current flood risk and bearing in mind the ongoing growth of value to be protected and the projected impacts of climate change, it was decided to change the Netherlands' perspectives on flood risk management towards a more flexible approach with a long-term horizon.

1.4. Aim of this paper: lessons learnt from the Dutch approach

This paper focuses on the recently changed perspectives in the Netherlands' risk reduction philosophy, aimed at protecting a densely populated delta at a lasting high safety level throughout this century while addressing the impacts of climate change and socio-economic developments. The changed perspectives reflect a shift from policy development and implementation following a disaster (reactive) to a nationwide approach of climate-proofing the delta and thus anticipating projections of climate change and socio-economic developments at the end of this century (proactive: prevention). In addition to presenting these changed perspectives, lessons learnt are presented from Dutch experience, which can be applied to flood-prone areas around the globe. The connecting theme through these lessons learnt is the realisation that a proactive approach to flood prevention instead of reacting to flood disasters pays off and should be the way forward for densely populated, flood-prone areas.

2. Origin of the Netherlands and its vulnerability to flooding

The relationship between the Dutch and 'their' water is founded in the country's genesis. In order to understand the Netherlands' flood risk policy, one needs to understand how the Netherlands came into being.

The Netherlands is a country by the North Sea, shaped by sediments from rivers and the sea, and by the heritage of peaty soils created by swamps some 3,000 years ago. Around AD800 people started to drain the swamps and cultivate the land. They dug the peat for fuel and to extract the salt. The drainage of the peat caused the land to subside, probably some three metres in a few centuries. The extraction of the peat made the situation even worse. More and more, the land became prey to the storm surges. The tidal inlets eroded and the sea penetrated further inland.

In the eleventh century the first dikes were built in an effort to stop the frequent floods from the sea. Storm surges at the end of the twelfth century turned a number of lakes into one big inland sea. To the north, the vegetation of large parts of the swamps along the coast was washed away and tidal flats were left behind. The land had to be drained for cultivation and this made the soft soil subside. This increased the need for building more, higher and stronger dikes. One thing led to another. During the eleventh and twelfth centuries rural communities decided to join forces. They connected their dikes and made large polders, and they made arrangements for the maintenance of these dikes. These arrangements were the first steps towards establishing the water boards that at present are still the key organisations for water management in the Netherlands. By around AD 1300 most of the marshland had been cultivated.

In the fifteenth and sixteenth centuries water ruled over land. This was only partly due to coastal floods. The water area also expanded because small lakes became bigger and bigger. These lakes started at the crossing of ditches or where people dug away the peat, mostly for fuel. Wind waves eroded the banks. The impact of storms became more dramatic as these lakes expanded. Since the peat was dug close to the major towns, the lakes formed close to these towns. They became a threat to these towns and it was decided to reclaim these lakes. In the next centuries over a hundred large lakes were reclaimed. Over recent centuries cities and towns have been built on the bottom of these lakes, up to nearly seven metres below mean sea level.

Currently, 55% of the Netherlands' land area is embanked, including 62% of its urban area with 67% of the Dutch population (De Moel *et al.*, 2011). These embanked areas are represented by the blue and green parts in Figure 2. In this figure the former lakes that have been reclaimed since the beginning of the seventeenth century can be seen clearly as the blue colours: these lakes are the deepest polders.

3. The Netherlands' flood protection policy since the coastal flood of 1953

3.1. The flood of 1953 and the 'first' Delta Programme

The most recent flood that hit the Netherlands was the flood of 1953. A storm pushed the water of the North Sea into the funnel-shaped southern bight of this sea between England and the European continent. Dikes were breached at a few hundred locations. Almost two thousand people drowned in the south-western part of the Netherlands. Hundreds of people lost their lives in Belgium and England as well. Almost 8% of the territory of the Netherlands became inundated. As a response, the Deltaplan (the 'first' Delta Programme), consisting of several dams and storm-surge barriers and the large-scale improvements of dikes, was carried out. The Deltaplan was completed in 1997.

3.2. The Netherlands' flood protection policy since 1953

Since 1953 no floods from the sea side have occurred in the Netherlands. In the river areas some very high flood levels were reached in the years 1993, 1995 and 1998. In some cases people needed to be evacuated, but the damage remained small and no human lives were lost.

In 1960 a Delta Commission proposed safety standards for flood defences along the Dutch coast and in the estuaries. At a later stage standards for river dikes were proposed. In 1996 these standards were

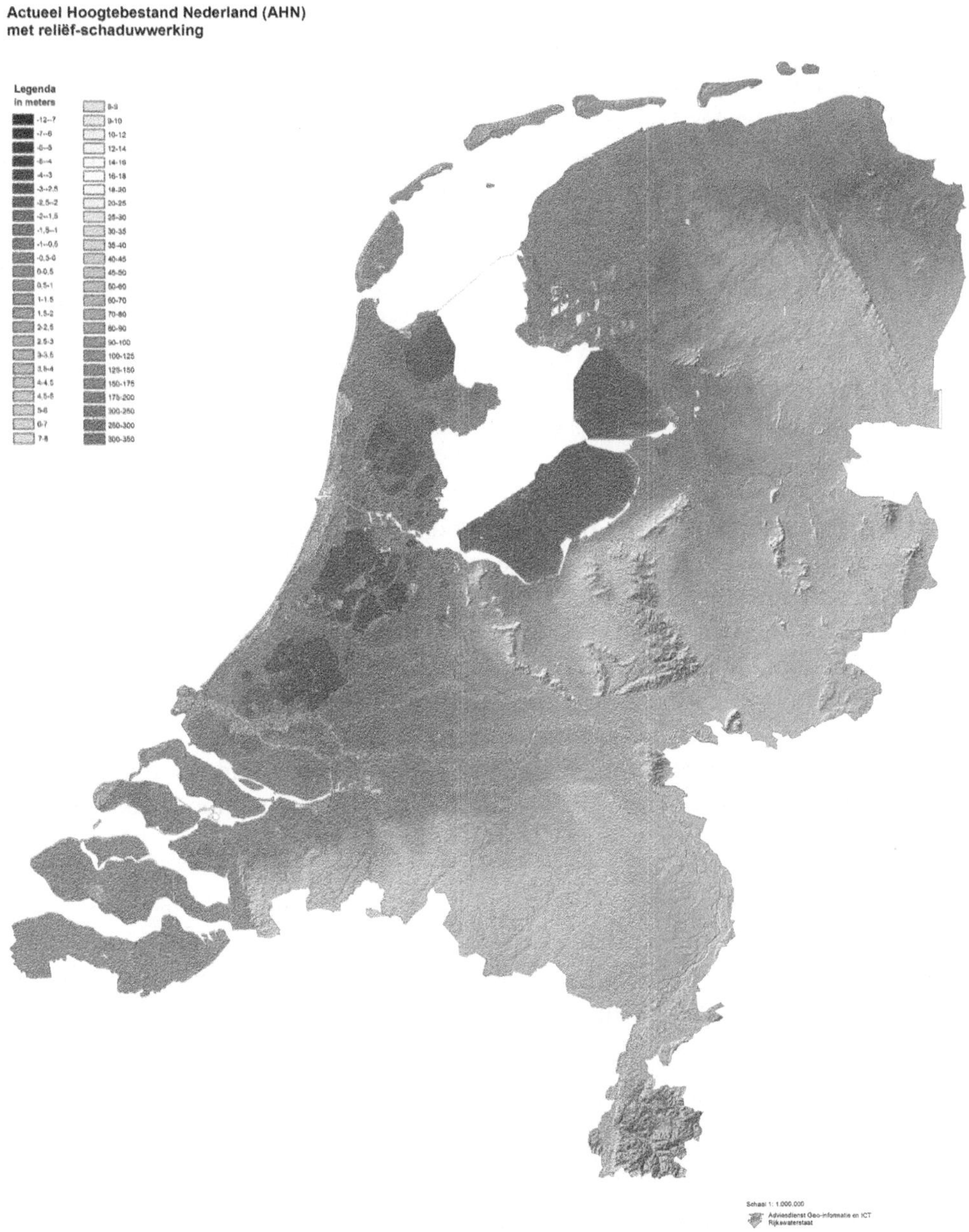

Fig. 2. Height of the Netherlands' surface area, showing the flood-prone (and embanked) parts in the west and the north including the deepest polders (blue) reclaimed from lakes and the sea (from: www.ahn.nl). The colour version of this figure is available online at http://www.iwaponline.com/wp/toc.htm.

anchored into law. The standards demand a minimum height and strength of dikes and constructions surrounding a given area, thus protecting this area from flooding from the sea, from the main rivers or from large lakes. Such an enclosed area protected by one set of dikes is called a dike ring. The flood-prone part of the Netherlands consists of 53 dike rings (and a number of small embankments along the River Meuse). The dikes are managed by so-called 'water boards', democratic bodies solely responsible for water management in certain territories.

The safety standards reflect the probability of occurrence of the highest water levels that need to be safely contained by flood defences.

The discussion on safety standards and on the Dutch flood protection policy took a new direction in 2005. This came about not as a result of actual flood problems in the Netherlands, but rather was triggered by Hurricane Katrina in Louisiana. The size and impact of the New Orleans disaster made the Dutch government realise that absolute safety or zero flood risk does not exist and that there was a clear need to review our flood policies and to develop a new forward-looking approach.

4. Socio-economic developments and projected impacts of climate change

4.1. Socio-economic developments

The IPCC (2014) reported a worldwide trend of flood risk increase with respect to coastal and river floods due to an increase of the vulnerability of people and their belongings along rivers and near the coast. This trend also holds for the Netherlands. During the twentieth century the amount of urban area in the flood-prone part of the Dutch delta (river area + coastal zone) increased about six-fold. Future projections of socio-economic changes show a further increase of urban area in the flood-prone zone by the year 2100 by 30% (scenario: low economic growth) up to 125% (scenario: high economic growth) with respect to the year 2000 (De Moel *et al.*, 2011).

A recent assessment of flood risk in the Netherlands, carried out within the framework of the Delta Programme (see below), has shown that 75% of the Netherlands' total flood risk is in the river area (Figure 3).

4.2. Sea-level rise

According to the most recent climate change scenarios for the Netherlands, sea-level rise near the Dutch coast in the period 2011–2070 will be 25–80 cm. It is not clear yet whether the storm surges

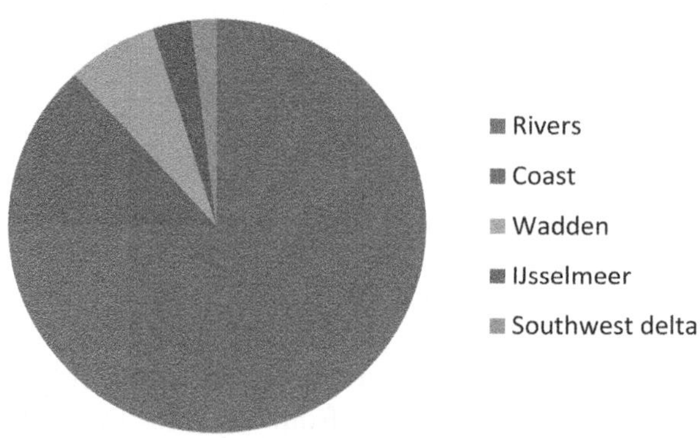

Fig. 3. The contribution of flood risk along the rivers, major lakes and the coast to total flood risk in the Netherlands according to an assessment of the Delta Programme (2014).

on the southern part of the North Sea will also increase (Koninklijk Nederlands Meteorologisch Instituut/Royal Netherlands Meteorological Institute (KNMI, 2014).

It is not the absolute sea-level rise but the relative sea-level rise (the rise with respect to the (subsiding) land) that determines changes in flood risk. Large parts of the flood-prone areas of the Netherlands are subsiding, mainly due to compaction and oxidation of the peaty soil, and due to the exploration of oil, gas and salt. Land subsidence in these areas will vary greatly from a few decimetres up to one metre between now and 2050 (Deltares, 2012).

4.3. River discharge

It is likely that the peak discharges of the main rivers in the Netherlands, the Rhine and Meuse, will increase over the course of this century (Drogue *et al.*, 2010; Görgen *et al.*, 2010). In 2016, when the Room for the River Programme is completed, the discharge capacity of the rivers Rhine and Meuse will be 16,000 and 3,800 m^3/s, respectively. This discharge capacity agrees with a peak discharge with a recurrence interval of 1,250 years. The Delta Programme (see below) anticipates a further increase in the discharge capacity of these rivers to compensate for the impacts of climate change to 17,000 and 18,000 m^3/s for the Rhine in 2050 and 2100, and to 4,200 and 4,600 m^3/s for the Meuse in 2050 and 2100.

5. The Deltaplan revisited: from flood protection to integrated risk management

The basic principles of the current Dutch flood risk policy are:

1. a minimum safety level for each citizen in the Netherlands (a probability to die due to a flood of at most 1/100,000 per year);
2. the introduction of a new set of safety standards for the Dutch flood defences based on a societal cost–benefit analysis;
3. counteracting social disruption in case of flooding;
4. protecting vital and vulnerable infrastructure.

These principles have been anchored in the approach of the Delta Programme.

5.1. Delta Programme approach

In 2010 the Delta Programme was set up, headed by the so-called Delta Commissioner. The main objective of the Delta Programme is to create a safe and attractive Netherlands, now and in the future, by providing adequate flood risk management and fresh water supply. The Delta Programme is a national programme, in which national, regional and local authorities prepare key decisions, develop strategies and implement measures, in close cooperation with the public, stakeholders and knowledge institutions (Van Alphen, 2014). Since 2013, a Delta Fund of about €1 billion per year will provide stability in financial resources to implement the decisions of the Delta Programme. Since 2012, the Delta Act has formed the legal basis for the implementation of the programme, the Delta Commissioner and the Delta Fund.

In 2014, after four years of analyses and strategy development, five key decisions have been presented to Parliament: three national policy frameworks (on flood risk management, on fresh water supply, and on water-proof and climate-proof reconstruction and new developments of the built environment), and two overall strategies for areas where flood risk and fresh water supply interact (the transitional areas between the rivers and the North Sea: Lake IJssel and the Rhine–Meuse estuary). Measures in the latter two areas affect the surrounding regions or impose boundary conditions on upstream regions.

These key decisions form a framework for regional strategies regarding six specific hydraulic regions in the Netherlands: the Coastal Area, The Wadden Sea, the south-western part with its estuaries, tidal inlets and lakes, the Rotterdam–Dordrecht tidal rivers, the upper stretches of the River Meuse, and the Rhine branches.

These key decisions and regional strategies have been developed with a long-term perspective, that is, a time horizon up to 2100. This long-term perspective stimulates the combination of investment agendas of different policy fields or authorities. In addition, it helps to anticipate climate change gradually by making future-proof decisions on nearby capital investments in infrastructure, flood defences and the built environment. On the other hand, this long-term perspective introduces uncertainty about the future conditions for which these measures have to be designed (Van Alphen, 2014). This uncertainty has been tackled by using four so-called Delta Scenarios that present the 'corner flags of the playing field of plausible futures' (Bruggeman & Dammers, 2013). Each scenario describes a plausible future in which climate change (rapid or moderate) and socio-economic development (growth or decline) are combined. The climate change parameters are based on KNMI (2014). The socio-economic parameters describe the future size and spatial distribution of population and land use, and constitute basic data for flood risk potential and fresh water demand.

The most important features of the Delta Programme are:

- the Programme is based on key success factors such as being led by an independent coordinator (the Delta Commissioner), having a long-term financial commitment of €1 billion/year and having a firm legal basis;
- many stakeholders are involved in a joint decision-making process to make sure that the decisions are socially broadly based;
- a risk-based perspective is taken as a basic principle;
- a flexible approach is adopted in the possible strategies by valuing flexibility with regard to the timing of implementation;
- various investment agendas are linked, and opportunities to mainstream actions with planned investments are used.

5.2. Building with nature

Ecosystem-based approaches are a cornerstone of disaster risk reduction (DRR) (IPCC, 2012). Environment degradation is a major driver of increased risk. Along many parts of the world's coasts, mangrove forests, for example, provide a natural barrier against extreme weather events and erosion. In addition, mangroves serve as feeding and nursery habitats for valuable fish species and sequester carbon more effectively than terrestrial forests.

In the Netherlands, ecosystem-based approaches are being applied more and more, where appropriate. Examples of these so-called 'building with nature' measures include sand nourishments along the coast

(Figure 4) and creating shallows or vegetation in front of flood defences to reduce wave current-induced erosion. The basic principle for coastal flood protection in the Netherlands is 'soft measures where appropriate, hard infrastructural solutions where needed'. The ecosystem-based approach leaves room for innovative solutions, such as the so-called sand engine, an island of sand nourished in connection to the coast and allowed (and meant) to erode and thus supply sand to most of the Dutch coast for years to come (Figure 5).

5.3. Adaptive delta management, adaptation pathways and tipping points

The Delta Programme comprises a cohesive set of projects (measures) for the short term, but also looks ahead to the medium and long term (up to 2050). Short-term measures must be logical in the

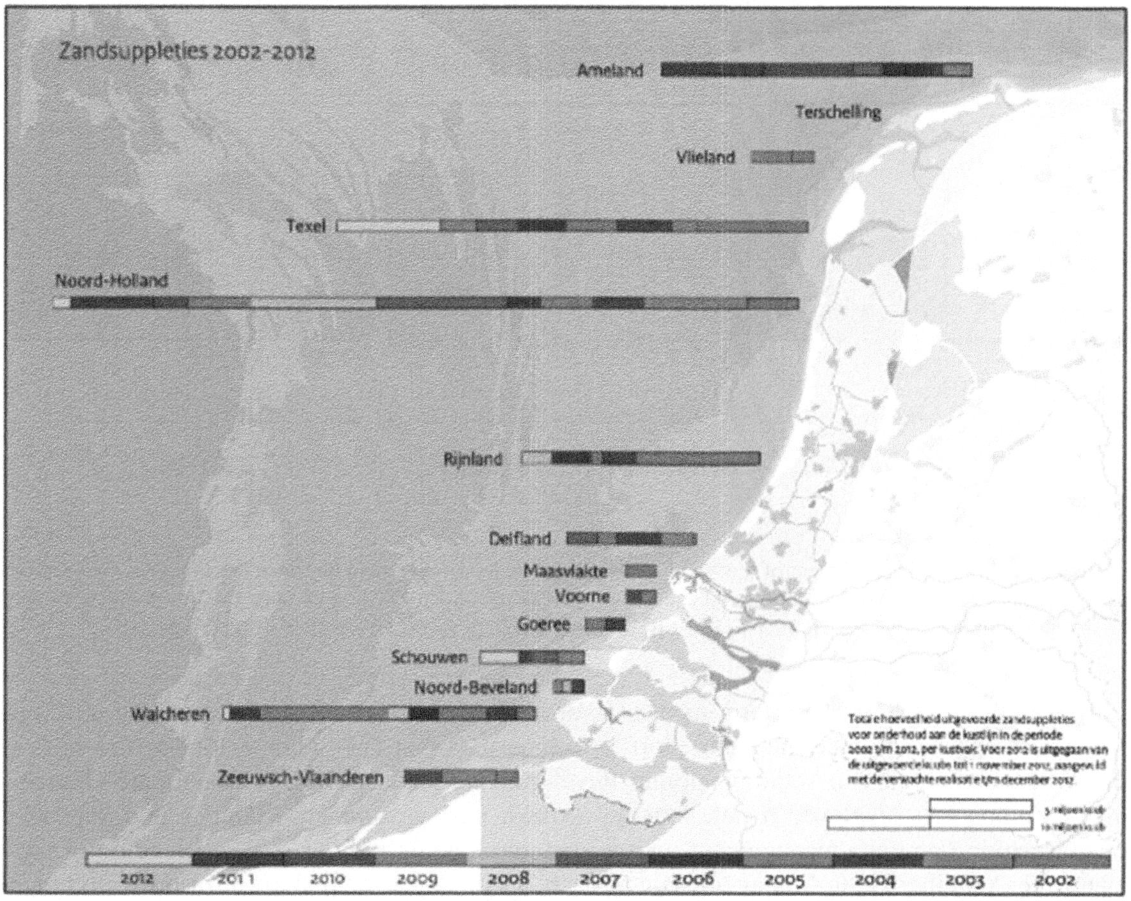

***Kaart 4** Uitgevoerde zandsuppleties in de periode 2002-2012 voor onderhoud van de kust.*
Bron: Deltaprogramma (2013) Nationale Visie Kust

Fig. 4. In the Netherlands the so-called 'building with nature' measures include sand nourishments along the coast. This map shows the nourishments that have been carried out during 2002–2012 at different parts along the coast.

Fig. 5. The innovative sand nourishment of the 'sand engine', where an island of sand has been nourished in connection to the coast and is meant to erode and thus supply sand to most of the Dutch coast for years to come. The development of the 'sand engine' is being monitored closely in order to learn from this innovative approach (*Source:* Rijkswaterstaat).

long term, must not obstruct long-term measures, or may even be necessary to keep long-term options open. It is complicated to take the right short-term measures that also address long-term issues while dealing with uncertainties in socio-economic developments and the impact of climate change. The Delta Programme has developed a new strategy to take into account uncertainties and dependencies in decision-making on delta management with a view to reducing the risk of overspending or underinvestment: the Adaptive Delta Management approach. In this approach, short-term decisions on water management, land use and spatial planning are linked to long-term issues on flood protection and freshwater supply, allowing switching between strategies through adaptation pathways (Figure 6).

Adaptation pathways are successions of strategies into the future in a changing environment. These pathways can be explored from many possible transient scenarios of climate and socio-economic developments. This approach shows the range of options from which policy-makers can choose. The

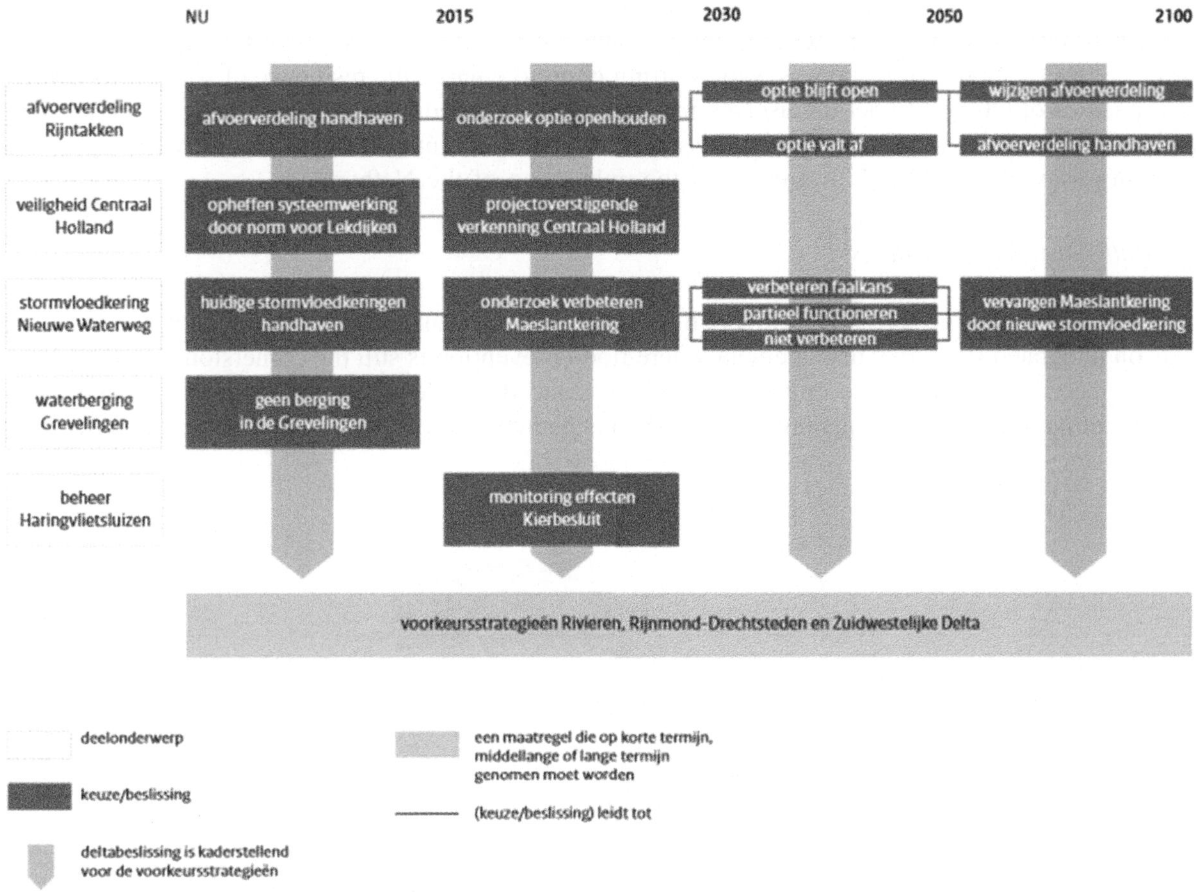

Fig. 6. An example of the approach of adaptation pathways that is used for decision-making within the Delta Programme. This example refers to the lower reaches of the rivers Rhine and Meuse near Rotterdam. A distinction is made between the decisions that have to be taken now (left-hand side) and that might be needed in the future (right-hand side), depending on future developments of climate change and socio-economic developments (*Source:* Delta Programme, 2014).

adaptation-pathways approach thus supports decision-making for sustainable water management in a changing environment (Deltares, 2010).

A switch to a new strategy is needed once an adaptation tipping point is in sight. Adaptation tipping points can be defined as points where the magnitude of change due to socio-economic developments, climate change or sea-level rise is such that the current strategy will no longer be able to meet the objectives (Deltares, 2010). At that time, the chosen strategy fails and alternative strategies are needed. An example of a tipping point is the functioning of current storm-surge barriers in view of sea-level rise.

5.4. Participation of citizens

As mentioned before, an important feature of the Delta Programme is the involvement of many stakeholders in a joint decision-making process. The citizens of the Netherlands are part of these stakeholders. It is difficult, however, to include citizens unless decisions affect their backyard. Flood risk awareness in the Netherlands is low and, according to a recent OECD report on water governance in the Netherlands, 'Dutch citizens tend to ignore water risks and functions when they develop property' (OECD, 2014). Flood risk awareness is the starting point to enlarge the resilience of society, to improve the capacities of citizens to act themselves in preparing for, and during, a flood, and thus avoid social disruption due to flooding as much as possible. Dutch water authorities, therefore, will continue to invest in raising the flood risk awareness of the population of the Netherlands.

5.5. Multi-layer safety strategy

The Delta Programme is a shift from a policy based almost exclusively on flood prevention to a wider focus on integrated flood risk management where flood prevention is still the cornerstone (layer 1) but in addition the vulnerability to the impact of floods is reduced by adequate spatial planning (layer 2) and more attention to contingency planning and crisis management (layer 3) (Figure 7). The extent to which

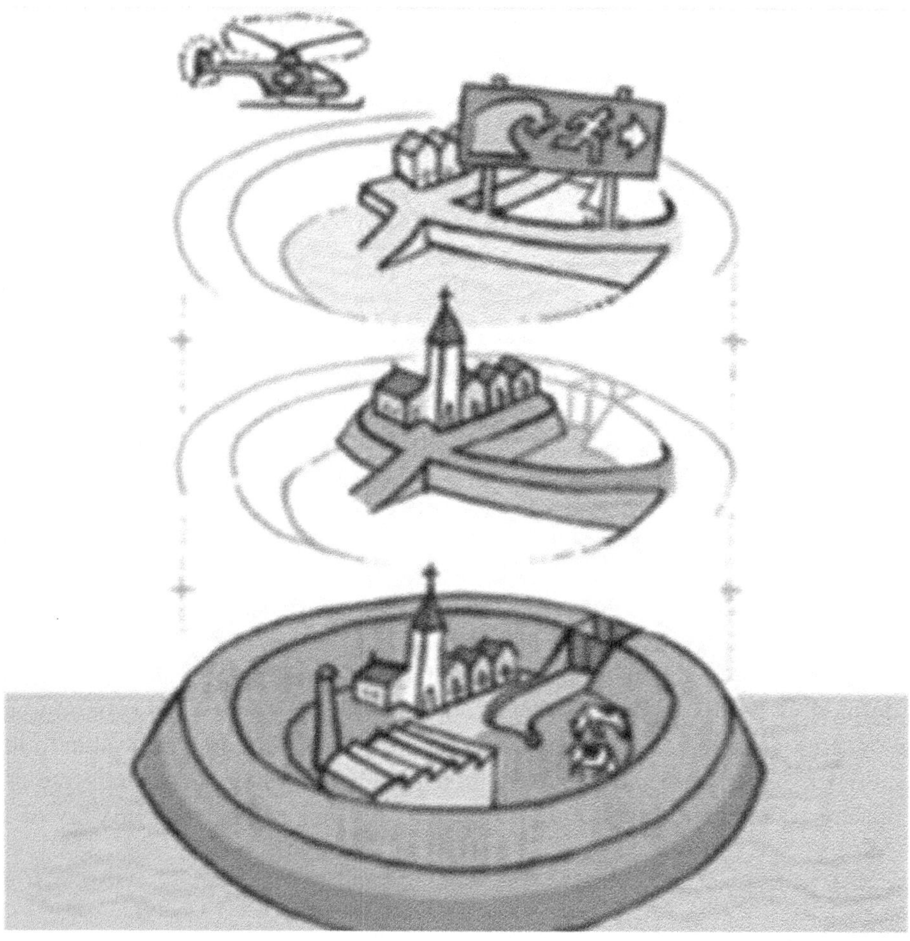

Fig. 7. The Delta Programme is a shift to a wider focus on integrated flood risk management where flood prevention is still the cornerstone (layer 1, below) but in addition the vulnerability to the impact of floods is reduced by adequate spatial planning (layer 2, middle) and more attention to contingency planning and crisis management (layer 3, above).

this approach of a multi-layer safety strategy can be applied varies from one part of the country to another. A lot can be learnt from cities elsewhere, such as London and Hamburg.

5.6. Implications for (multi-level) governance

The changing perspectives in the Netherlands' flood risk reduction philosophy call for changing perspectives in water governance. Capacities are needed to take decisions under uncertainty, to be flexible and to be able to switch to different strategies when needed, and to think and act across the boundaries of sectors or interests. In the words of the IPCC (2012):

> *'Adaptive capacity underlies action and is defined in this report as the combination of strengths, attributes, and resources available to an individual, community, society, or organization that can be used to prepare for and undertake adaptation. Adaptive capacity can also be described as the capability for innovation and anticipation, and the capacity to generate experience in dealing with change, the ability to learn from mistakes.'*

One of the adaptive capacities is fair governance. This includes the elements of accountability, responsiveness, equity, and legitimacy (Grothmann *et al.*, 2003). When taking measures to reduce risks, these elements should be clearly addressed. Accountability is addressed by designating the responsibilities of all stakeholders that cover different parts of risk management. In the multi-layer safety strategy, for instance, the state sets the safety standards for primary flood defences; regional water boards are responsible for dike maintenance; provinces direct spatial planning; and municipalities, police, fire fighters and health authorities are responsible for contingency planning and crisis management. In the Delta Programme, responsiveness is addressed through a trajectory of stakeholder participation. Equity is reflected in, among others, the principle that people pay tax for regional water management (including dike maintenance) in the area of their water board. Legitimacy is secured by reporting the process of the Delta Programme to Parliament each year and by anchoring the Programme and its key decisions in the Delta Act.

5.7. Low probabilities – high consequences

In the past decades, focus on flood risk management in the Netherlands was almost exclusively on flood protection. This resulted in the world's finest flood defence with the highest safety standards. The other side of this policy is the fact that options to deal with the consequences of a flood in case (one of) the dikes should fail (shelters, evacuation strategy, etc.) were hardly addressed. One of the perspectives in the Netherlands' flood risk reduction philosophy that has changed is the attitude towards flood disasters of low probabilities but with high consequences. This change is reflected in the policy goal to counteract social disruption in case of flooding. The goal to be advanced is a society that is more resilient. The higher the level of resilience, the lower the level of disruption by flooding.

The options to reduce social disruption in case of flooding have been assessed by the Netherlands Environmental Assessment Agency (PBL, 2014). It was concluded that social disruption can be reduced by:

- taking concrete measures to restrict the number of fatalities during a flood, such as building dikes too wide to fail (and thus excluding the possibility of being flooded by surprise), creating flood shelters and designing an effective evacuation strategy;

- acquiring a good understanding of the extent to which vital functions (such as energy and drinking water supply) can be secured;
- investigating the possible contribution of recovery plans to the reduction of social disruption;
- organising a communication with society on current flood risk, flood risk policy and disaster management.

6. Dutch national risk assessment: flood risk and other risks compared

In 2007 a national safety and security strategy was introduced in the Netherlands, based on a national risk assessment of all major threats and hazards that might strike the country, including floods and droughts. This assessment comprises an estimation of the probability and possible impact of a threat or hazard; the estimate of the impact includes both physical (economic damage, number of casualties, etc.) and social criteria (disruption to everyday life, loss of trust in authorities, etc.). By using the same method for all threats and hazards, the resulting impacts of these threats and hazards can be compared, and the results can thus be used for decision-making on investments in overall risk reduction at a national scale (Mennen & Van Tuyll, 2014).

Since 2012 the Dutch steering committee for national safety and security has acted as the national platform for DRR. Results are positive. Understanding of national risks has increased, capacities for DRR were enhanced, and the (intergovernmental and public–private) network for DRR was strengthened.

7. Lessons learnt for flood-prone areas around the globe

In the Netherlands in recent years an integrated, adaptive approach has been developed based on the recognition that risks, if not properly addressed in advance, may lead to the occurrence of disasters. Facing risks, on the other hand, creates an opportunity not only to avoid a risk turning into a disaster, but also to use the threat as an opportunity to transform society to a higher level of sustainability (Takeuchi & Chavoshian, 2013).

Examples of flood disasters can be found on a regular basis, even in Western Europe. The changing perspectives in the Netherlands' flood risk reduction philosophy illustrate that a widespread and effective implementation of an integrated, adaptive approach to flood risk management requires a reframing of the paradigm of disaster risk management from its traditional focus on mitigating the (direct) impacts of disasters using stand-alone and ad hoc interventions to a broader focus on prevention, mitigation, preparedness and vulnerability reduction. This broader focus includes flexibility to switch to other strategies when needed, in view of the uncertainties of future socio-economic developments and climate change. There is an element of experimentation and learning in this. Besides, this broader focus calls for changing perspectives in water governance as well, with responsibilities at all levels: globally, nationally, as well as locally and at community level.

The context of the Netherlands may be unique in view of its long-term experience in dealing with floods for almost a millennium, a large part of its territory being flood-prone (and partly below sea level) and densely populated, and its wealth that allows for building and maintaining the 'world's finest flood defence'. The principles behind the Netherlands' flood risk reduction philosophy can be applied universally, however. Three basic principles are as follows:

There is a sense of urgency now to act on water-related disasters. Climate change is exacerbating the extremes in hydro-meteorological events. Together with other global drivers under change – population growth, rapid urbanisation, increased asset values – this will increase the frequency and impact of water-related disasters in a number of regions around the globe.

We must invest in better understanding the problem and adopt a proactive yet flexible approach. More data and better tools for risk assessment are ready for use now and need to be more widely deployed to identify and prioritise actions. A vulnerability assessment provides an essential information tool in water-related disaster management. Uncertainties are no excuse for inaction: uncertainties are inherent in long-term planning and should be accounted for in a comprehensive, flexible and adaptive approach.

We must take preventive measures before disasters strike. Risk reduction, preparation and prevention are sensible investments that pay off in terms of reduced loss of life, the avoidance of damage, and long-term economic growth and stability. Risk prevention should be integrated with long-term planning. This allows communities and decision-makers to identify and exploit opportunities for synergies with planned investments, including plans for adaptation to climate change.

From these principles the Dutch approach may serve as an example of an effective approach to take integrated flood risk management to a higher level: create a framework of legitimacy (the Dutch Delta Act); explore the future (by using scenarios); draw up a vision to deal with the outcomes of these scenarios; develop a set of measures along adaptive pathways; secure funding for decades ahead (the Dutch Delta Fund); and act smart when implementing these measures (by looking for combinations with other spatial measures, such as building a car park combined with a dike that are both subsequently covered by a dune (Figure 8)).

Fig. 8. Near the village of Katwijk the coastal flood defence needs to be strengthened and this opportunity is being exploited by combining dike construction with building a car park and subsequently covering both of them underneath a dune.

With respect to disaster management, we must be particularly aware of the impact of disasters to vulnerable groups and of the need to tailor policies to improve people's resilience. Innovative technologies and instruments will become available to support disaster risk management, such as information and communication technologies, early-warning systems, resilient infrastructure and buildings, green infrastructure, climate and integrated disaster risk modelling, and ecosystem-based approaches. This will also lead to increased business opportunities and contribute to green growth.

8. From concepts to action

The example of the changing perspectives in the Netherlands' flood risk reduction philosophy illustrates the concepts of integrated DRR to water-related disasters in general. Also with respect to droughts, for instance, measures can be designed based on future scenarios and a flexible, adaptive approach that allows for switching between strategies along adaptation pathways when needed in view of socio-economic developments or climate change. Steps must be taken to turn these concepts into action around the globe:

1. Disaster risk reduction, water resources' management and climate adaptation should no longer be treated as separate topics but should be merged into an integrated approach. Spatial planning plays a central role in this. The close link between DRR and climate adaptation has already been stressed by the IPCC (2012), for instance by concluding that *'Disaster risk management and adaptation to climate change can reduce exposure and vulnerability to weather and climate events and thus reduce disaster risk, as well as increase resilience to the risks that cannot be eliminated.'* This integrated approach can contribute to sustainable development and economic growth.
2. Current international processes should be used to get things done. This calls for a further elaboration of the Hyogo framework, the Sustainable Development Goals (including SDGs on water and urbanisation), the United Nations Framework Convention on Climate Change (UNFCCC), and the Third United Nations Conference on Housing and Sustainable Urban Development (Habitat III).
3. Make sure that the boundary conditions for the implementation of the necessary measures to reduce risk and increase resilience are in place, such as long-term funding, the elements of fair governance, and stakeholder participation.

References

Bruggeman, W. & Dammers, E. (2013). Deltascenarios for 2015 en 2100, p. 65.

Conradt, T., Roers, R., Schröter, K., Elmer, F., Hoffmann, P., Koch, H., Hattermann, F. F. & Wechsung, F. (2013). Vergleich der Extremhochwässer 2002 und 2013 im deutschen Teil. *Hydrologie und Wasserbewirtschaftung* 57, 241–245.

Cour des Comptes (2012). *Summary of the Public Thematic Report by the Cour des Comptes*, p. 31.

Delta Programme (2014). *Work on the Delta. Favourable Solutions for Questions and Ambitions*. Delta Programme Report. Ministerie van Infrastructuur en Milieu, Den Haag, The Netherlands, p. 132 [in Dutch].

Deltares (2010). *Sustainable Water Management Under Climate Change*. Deltares, Delft, The Netherlands, p. 20.

Deltares (2012). *Fresh Water Supply in the Netherlands. An Analysis of National Bottlenecks in the 21st Century*. Deltares Report 1205970-000. Deltares, Delft, The Netherlands p. 292 [in Dutch].

De Moel, H., Aerts, J. C. J. H. & Koomen, E. (2011). Development of flood exposure in the Netherlands during the 20th and 21st century. *Global Environmental Change 21*, 620–627.

DETEC (2008). *The Floods of 2005 in Switzerland.* Synthesis Report on the Event Analysis. Federal Department for the Environment, Transport, Energy and Communications, Switzerland, p. 23.

Drogue, G., Fournier, M., Bauwens, A., Commeaux, F., De Keizer, O., Francois, D. & Guilmin, E. (2010). *Analysis of Climate Change, High-flows and Low-flows Scenarios on the Meuse Basin.* AMICE WP1 Report – Action 3. Amice Meuse Maas, Liege, Belgium, p. 67.

European Environment Agency (EEA), JRC & WHO (2008). *Impact of Europe's Changing Climate – 2008 Indicator-Based Assessment.* EEA Report No. 4/2008 – JRC Reference Report No. JRC47756 European Environment Agency, Copenhagen, Denmark, p. 246.

Görgen, K., Beersma, J., Brahmer, G., Buiteveld, H., Carambia, M., De Keizer, O., Krahe, P., Nilson, E., Perrin, C., Lammersen, R. & Volken, D. (2010). *Assessment of Climate Change Impacts on Discharge in the Rhine River Basin: Results of the RheinBlick2050 Project*, CHR rapport I-23. Commission for the Hydrology of the Rhine, Lelystad, The Netherlands, p. 229.

Grothmann, T., Grecksch, K., Winges, M. & Siebenhüner, B. (2013). Assessing institutional capacities to adapt to climate change: integrating psychological dimensions in the adaptive capacity wheel. *Natural Hazards and Earth System Sciences, 13* (12), 3369–3384. doi:10.5194/nhess-13-3369-2013.

Habersack, H. & Krapesch, G. (2006). *Hochwasser 2005 – Ereignisdokumentation der Bundeswasserbauverwaltung, des Forsttechnischen Dienstes für Wildbach- und Lawinenverbauung und des Hydrographischen Dienstes.* Universitat fur Bodenkultur, Wien, Austria, p. 150.

IPCC (2012). *Managing the Risks of Extreme Events and Disasters to Advance Climate Change Adaptation.* A Special Report of Working Groups I and II of the Intergovernmental Panel on Climate Change. Cambridge University Press, Cambridge, United Kingdom, p. 582.

IPCC (2014). *Climate Change 2014: Impacts, Adaptation, and Vulnerability.* Working Group II Contribution to the Fifth Assessment Report of the Intergovernmental Panel on Climate Change. Cambridge University Press, Cambridge, United Kingdom. http://www.ipcc.ch/ (accessed October 2014).

KNMI (2014). *KNMI'14 Climate Scenarios for the Netherlands.* KNMI, de Bilt Netherlands, p. 34 [in Dutch].

Kolen, B. (2013). Certainty of Uncertainty in Evacuation for Threat Driven Response. PhD Thesis, Radboud University Nijmegen, The Netherlands, p. 315.

Mennen, M. G. & Van Tuyll, M. C. (2014). Dealing with future risks in the Netherlands: the national security strategy and the national risk assessment. *Journal of Risk Research.* RIVM, Bilthoven, The Netherlands. http://dx.doi.org/10.1080/13669877.2014.923028.

OECD (2014). *Water Governance in the Netherlands: Fit for the Future? OECD Studies on Water.* OECD Paris, France. http://dx.doi.org/10.1787/9789264102637-en.

PBL (2014). *Low Probabilities – High Consequences. Casualties and Social Disruption As a Focus for Flood Safety Policy.* PBL Report. Netherlands Environment Assessment Agency, The Hague, The Netherlands p. 93 [in Dutch].

Pitt Review Team (2008). *Learning Lessons from the 2007 Floods*, p. 462.

Pollner, J., Kryspin-Watson, J. & Nieuwejaar, S. (2008). *Disaster Risk Management and Climate Change Adaptation in Europe and Central Asia.* Report World Bank. Global Facility for Disaster Reduction and Recovery, Washington, DC, USA p. 54.

Săvoiu, G. (2008). *Adaptive Management of the Climate Change Problem: Bridging the Gap Between Research and Public Policy. Munich Personal RePEc Archive*, MPRA Paper No. 13564. Annals of the Oradea University, Fascicle of Management and Technological Engineering, Vol. 7 (17), pp. 2733–2742 (http://mpra.ub.uni-muenchen.de/13564/).

Takeuchi, K. & Chavoshian, A. (2013). *Floods: From Risk to Opportunity.* Vol. *357*, *IAHS Publ.*, IAHS, France, pp. 3–9.

Ten Brinke, W. B. M., Bannink, B. A. & Ligtvoet, W. (2008). The evaluation of flood risk policy in the Netherlands. *Journal on Water Management 151* (4), 181–188.

Ulbrich, U., Brücher, T., Fink, A. H., Leckebusch, G. C., Krüger, A. & Pinto, J. G. (2003). The central European floods of August 2002: Part 1 – Rainfall periods and flood development. *Weather 58*, 371–377.

Van Alphen, J. (2014). The Delta Programme and Updated Flood Risk Management Policies in The Netherlands. In: *Proceedings of the 6th International Conference on Flood Management*, Brazil, 2014. Brasilian Water Resource Association, Brasil.

Visser, H., Petersen, A. C. & Ligtvoet, W. (2014). On the relation between weather-related disaster impacts, vulnerability and climate change. *Climatic Change 125*, 461–477.

Wisner, B., Blaikie, P., Cannon, T. & Davis, I. (2005). *At Risk (Second Edition). Natural Hazards, People's Vulnerability and Disasters.* Routledge, London, p. 471.

Water-related disaster risk reduction (DRR) management in the United States: floods and storm surges

Jerome Delli Priscoli[a] and Eugene Stakhiv[b]

[a]*Corresponding author. USACE Institute for Water Resources, World Water Council and School of Advanced International Studies (SAIS), Johns Hopkins University, Washington, DC, USA*
E-mail: priscoli@erols.com

[b]*International Center for Integrated Water Resources Management (ICIWaRM), USACE Institute for Water Resources and John Hopkins University, Washington, DC, USA*

Abstract

Water resources planning and management has evolved in the United States through several distinct stages over the past two centuries, transitioning from a concern for inland waterways transportation to single purpose flood control and finally to multiple purpose large reservoirs. Disaster risk reduction (DRR) was always considered to be one of the main goals of these strategies. Reviewing history, this paper describes a US federal system that presents major challenges to coordinating water resources development and DRR, at both the watershed and metropolitan area scales. The paper reviews the performance of existing flood protection systems of three recent disasters. Federal, state and local responses to these major events have been mixed, as regulatory and management agencies with different evaluation frameworks and decision rules attempt to coordinate their respective responses. The cases revealed new vulnerabilities and weaknesses in the US DRR responses and planning, while contrasting the relative successes of long-term, strategic DRR planning and investments in the case of the Mississippi River and Tributaries (MR&T) system. The paper analyzes this history and recent cases primarily from the perspective of the US Army Corps of Engineers.

Keywords: Disaster Risk Reduction; Floods; Katrina; Mississippi; Storm surges; Superstorm Sandy

1. Introduction

Flood and water disaster risk reduction (DDR) management in the United Sates is carried out in the context of a federal system (Kenney, 1997). There has been over a century of United States Army Corps of Engineers (USACE) and federal attention to flood risk management. Within the US federal system, all powers not ascribed to the federal government in the National Constitution remain with sovereign states. This is important in water resources since states retain sovereignty over water within their

doi: 10.2166/wp.2015.004

territories. Consequently, the federal role in water resources development evolved over US history for specific reasons in response to specific needs (Holmes, 1972).

Traditionally, there has been state and local resistance to federal government actions which are often seen as interference in water and natural resources management over which states have first sovereignty. Federal actions have been taken to achieve limited coordination on water resources for specific purposes such as assuring common standards, assuring interstate commerce and dealing with interstate conflicts. In addition, a strong tradition of property rights, individualism, markets and private ownership exists in the USA. There is also general resistance to land use planning (especially if it is from the national government) which makes water-related DDR management difficult, since most viable solutions that contribute to infrastructure resilience must integrate current and future land uses.

A long standing culture of primarily structural responses, funded by the federal government, has been significantly modified with recent water-related disasters such as Katrina and Sandy. But when disasters do occur there is a long tradition of helping the victims. Recent post-disaster recovery responses, however, have exposed discontinuities among assistance, mitigation and prevention programs of various federal agencies in the overall flood and water disaster management picture (US Congressional Research Service (CRS), 2013). These discontinuities are due mainly to the growing difficulties of a fragmented federal governance system in dealing with large-scale disasters, and a move toward environmentally sustainable solutions (US Government, GAO, 2014).

The loss of life and property damage resulting from a series of major riverine and coastal flood events over the last decade in the USA has reminded the nation of the reality of low probability, high consequence flood events. Notable events include Hurricane Katrina, which caused great destruction in the Gulf Coast states in 2005, riverine flooding in the Midwest in 2008, extensive floods along the Mississippi and Missouri River basins in 2011, and storm surge flooding in the New York region caused by Superstorm Sandy in 2012. Meanwhile, many scientists and practitioners say that the likelihood of extreme flood events is increasing in unpredictable ways due to climate change and 'non-stationarity' (increased variability) of flood and drought events (Milly *et al.*, 2002).

What could not be foreseen at the beginning of the federal environmental era in 1969, was the proliferation and increasing power of federal regulatory agencies which had legislative mandates to protect the environment, endangered species, historical and cultural sites, water quality, wetlands, air quality, coastal zones and aquatic ecosystems. These were federal statutes that applied to all 50 states and which were increasingly used, through agency administrative interpretations, to shape water resources policies and solutions, including floodplain management (FPM) and protection of ecological values of floodplains.

US federal water management agencies advocated multiple objectives and multiple means for water management, with the overriding objective of promoting economic growth and development, and reducing losses from floods and droughts (Billington *et al.*, 2005). However, the new regulatory agencies have largely superseded these internally consistent planning paradigms of the federal water management agencies, created by the US Water Resources Council (1973, 1983), and have imposed significant but often inconsistent constraints on many of the solutions advocated by the public and the federal water management agencies, particularly those engaged with flood damage reduction and FPM. A variety of prescriptive environmental criteria and standards, not always consistent among the federal agencies, indirectly served as constraints on water resources management plans by substantially reducing the range of economically feasible options (Stakhiv & Major, 1997). These environmental criteria superseded the traditional purposes and procedures associated with water resources development solutions.

As the number of planning constraints grew (as part of vaguely defined sustainable solutions), so did the costs of solutions, making them less affordable for communities.

Figure 1 (US Congressional Research Service, 2008) depicts the evolution of the 'federal interest' as seen through federal expenditures on water resources from the 1950s to early 2000s. The National Environmental Policy Act (NEPA) was passed on 1 January 1970 causing funding for water quality investments to increase dramatically and then level off in the 1990s through to the present. However, the role of federal authorities in water resources dramatically shifted from the traditional large infrastructure investment and planning of the pre-NEPA era to one of operations and maintenance and environmental regulation as normative drivers of US water management. Today, the understanding and policy implications of transactions and opportunity costs of this shift are growing, and the USA is continuously seeking better institutional ways to incorporate the important environmental concerns into norms for water planning and investments, while maximizing the creative formation of alternatives and negotiating among the sovereign entities that have authority over water.

2. History of US water DRR policies

From the 18th to early 20th century, flood management and DRR was primarily dealt with as a local problem, with small-scale structures as the preferred approach to DRR. During the mid-20th century the national federal role grew and so did the use of large structures. Later in the 20th century the USA moved back to a more local focus. However, there is now a movement to combine a portfolio of small and large structures with non-structural responses into a new understanding of risk management, and especially residual risk and who bears that residual risk. One can track the changes in US orientation to managing water-related disasters through changes in the language used to describe its policy

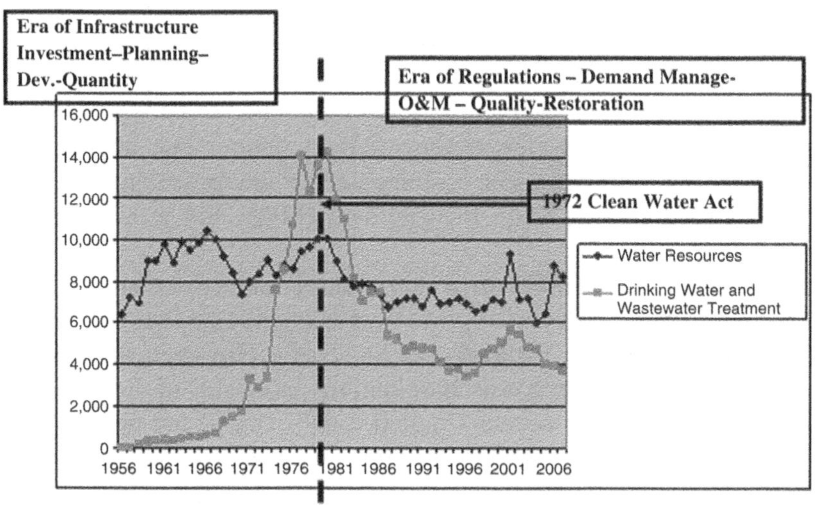

Source: CRS, with data from the Congressional Budget Office, CBO Infrastructure Spending Supplemental Tables, 2008 Update, available at http://www.cbo.gov/publication/19633?index=9135

Fig. 1. Federal water resources and water quality (1957–2007; in millions of 2006 dollars).

approaches: in the 1800s it was 'flood prevention'; in the early 1900s, 'flood control'; in the mid-1900s, 'flood reduction'; in the late 1900s, flood plain management (FPM); and more recently, 'DRR' (US Federal Task Force on Flood Control Policy, 1966; Holmes, 1972; Rosen & Reuss, 1988).

There has been a complex institutional history of FPM in the USA from 1900 to contemporary times. FPM has evolved out of several main streams of US resources policies over the years. These include resources protection, natural resources management, flood control, human resources management and disaster assistance. These policies have evolved within various US Federal Agencies such as the US Department of Agriculture's (USDA) national forest programs, early flood studies by US Geological Survey (USGS), National Park Service, and the Soil Conservation Service (SCS, now the Natural Resources Conservation Service). By the 1930s, federal funds were also supporting wildlife management. The Nature Conservancy was established, as was the Fish and Wildlife Service, Historic Preservation Act, the Wild and Scenic Rivers Act, and in the early 1970s the sweeping Clean Water Act (US Federal Task Force on Flood Control Policy, 1966).

The early Corps of Engineers authorizations for flood control, followed by the Bureau of Reclamation and the gradual extension of flood control beyond the Mississippi in the 1920s, were followed by states starting to regulate dams and encroachments. The sweeping Flood Control Act of 1936 set the stage for comprehensive, federally-sponsored flood damage reduction programs for the next 50 years. The Tennessee Valley Authority (TVA) was created as an instrument of regional economic development. In the 1940s and 1950s, there was an emergence of FPM ideas by Gilbert White (1945) and the University of Chicago. FPM services and flood way regulations eventually flowed into unified FPM programs such as the National Flood Insurance Act, NEPA (1970), and the establishment of the Environmental Protection Agency (EPA).

The evolution of disaster assistance shows less formal activity. The Red Cross was chartered to provide disaster relief for floods in the early part of the last century. However, it was not until the early 1950s that the emergence of the first comprehensive federal disaster relief programs was seen.

Today, these various policies and programs are being unified into FPM. The Unified National Program (UNP) is not a formal legislated program but one representing an evolving set of combined federal agency guiding principles. It is underlain by the implicit expectation that its implementation will be carried out largely through ongoing programs, activities and policies of the federal, state and local agencies, and professional groups with responsibility for, or interest in, flood-prone areas. FPM, based largely on the non-structural approaches advocated by Gilbert White (1945, 1969). It inherently relies on the land use regulatory powers of the state and local entities – not federal agencies.

Within the framework of the Unified National Program (US Federal Task Force on Flood Control Policy, 1966; US Federal Interagency Management Task Force, 1994), the intent was local governments and the private sector will set their own specific objectives, in coordination with their land use zoning authorities to both meet individual needs and also help achieve national goals. Each entity can lend additional momentum to the UNP by demonstrating exemplary FPM through its own actions. The federal government provides incentives to conform to the goals of the unified national FPM program, as well as regulatory penalties for non-conformance. It is the regulatory programs of the various federal agencies that have become the de-facto implementation and enforcement mechanisms of national FPM goals, as these programs control the vast array of private sector activities in the floodplains such as shopping malls, highways, waterfront renewal, urban renewal, parks, agricultural activities, storm water management, commercial navigation and water quality (Burby *et al.*, 1988).

There is an ever-growing number of private and public actors involved in water-related DRR. At the same time, there is no national water council or formal federal coordination mechanisms for rapidly developing, changing or implementing water policy. Current US water policy, including adaptation to climate change, is largely driven by the President's Executive Orders and implemented by the large federal agencies. However, executive orders, as opposed to legislation, may be and often are rescinded by subsequent presidents. Today, more than a dozen US federal agencies are involved in managing programs for flood DRR.

Much of the US history of flood management has been written around the lower Mississippi and the delta. In 1543 the Spanish explorer, Hernando de Soto noted how the Native Americans raised mounds by hand and built them high where they could. By 1727, Nouvelle Orleans was protected by a 4 foot (1.22 m) embankment. From 1823, a federal role in water resources began to emerge, consisting mostly of clearing navigation hazards, such as sandbars and debris, with continual Congressional debate about flood control roles by the federal government in response to a series of severe floods in the mid-1800s–1917 (Holmes, 1972; Arnold, 1988; Rosen & Reuss, 1988).

Debates on flood control in the Mississippi River system in the 1850s centered on the recommendations in 1852 of a private engineer, Charles Ellet, Jr., who was educated at L'Ecole Polytechnique in Paris. He advocated a mixed strategy of levees, river control structures and outlets for the lower Mississippi River. Severe floods in 1858 and 1859 swept away many of the private levee systems. At the same time, the Corps of Engineers commissioned Captain Andrew Humphreys and Lieutenant Henry Abbot to review the situation, and their detailed report of 1861, which was based on a comprehensive series of hydraulic measurements and a decade of research, advocated a 'levees only' strategy for the lower Mississippi River. The Humphreys view prevailed for 60 years until the devastating floods of 1917 and 1927. In the meantime, the Mississippi River Commission was founded in 1879 to coordinate and manage the various public and private flood control and navigation measures that were proliferating in the lower Mississippi valley (Holmes, 1972; Arnold, 1988; Rosen & Reuss, 1988).

After the flood disasters in 1916, Congress passed the Flood Control Act of 1917, wherein the federal government, for the first time, was directly charged with planning and financing flood control measures in the lower Mississippi and Sacramento rivers. The Act instituted cost-sharing between the federal government and local governments, where the federal government would pay two-thirds of the costs, and the local share would consist mostly of lands, easements and rights of way. The Flood Control Act of 1928 followed the 1927 flood disasters and established a comprehensive flood control strategy for the lower Mississippi River. Many call the 1927 floods the greatest disaster in US history, as 700,000 people were left homeless and there were $250 million (in 1927 US dollars) in economic losses. In the previous 200 years, locals had spent approximately $300 million in lower Mississippi flood works. However, in the single 1928 Act, Congress authorized $325 million for flood control works. This expenditure represented the greatest percentage of the US budget for water projects ever authorized in the USA (Holmes, 1972; Arnold, 1988; Rosen & Reuss, 1988).

The decisive turning point in a comprehensive federal role and responsibility for national flood control occurred after the devastating floods of 1936, which ravaged the Ohio River valley and the upper and lower Mississippi River. The Flood Control Act (1936) expanded the Corps of Engineers role, and stipulated that flood control reservoirs were to be a part of the solution set, and that federal investments were to be based on a benefit–cost analysis. The importance of upstream practices was clearly linked to downstream flooding. While the SCS of the Department of Agriculture was assigned to work on the upstream tributaries, the Corps was assigned responsibility for all the major rivers of the USA. A good part of the rationale for an elevated federal role stemmed from the need to overcome the devastating effects of the 1930s economic Depression, and to spur economic growth and provide employment for the millions out of work. The TVA and the Grand Coulee Dam were both started during this period as part of the broader social objectives of a federal public works program (Holmes, 1972; Arnold, 1988; Rosen & Reuss, 1988).

The TVA began a sweeping regional program of resource development that included the construction of multi-purpose dams, including flood control. The Bureau of Reclamation and the USDA began including flood control with other considerations in their projects. Floodplain lands, waters and natural resources received protection, if at all, only under a handful of laws and programs for preserving special areas such as public parks, forests and wetlands.

During this period, human and natural resources were managed separately. The federal government was the principal actor simply because it was the best organized, could tackle inter-jurisdictional problems, and marshal adequate personnel and financial resources. Throughout this era, states and local entities were not successful in controlling floodplain encroachments by the agricultural sector or municipalities. Despite increased effort in the physical control of flooding, economic damages continued to grow as the national economy expanded. FPM and restricting development in floodplains by the local entities was not considered a federal responsibility.

FPM was not a common term during the early and middle decades of the 20th century. Nevertheless, a number of diverse efforts were being undertaken that would later be recognized as precursors to contemporary FPM. The Corps of Engineers and the USGS continued to set up stream gage networks during this period. The US Forest Service studied the relationship between timber harvest and runoff. A national program for the study and management of upstream watersheds was authorized. The US SCS helped individual landowners and operators apply conservation measures (including flood prevention) in 2,600 soil conservation districts.

A few states adopted limited FPM regulations, mostly to prevent channel encroachments. Localities and special districts continued to build levees, and modify and maintain small channels. In 1953, the TVA initiated a comprehensive regional FPM program and during the early 1960s the Corps began offering FPM services nationally (Holmes, 1972; Arnold, 1988; Rosen & Reuss, 1988).

Between 1936 and 1952, the US federal government spent billions of dollars on flood control projects and storage for single and multi-purpose projects. During this time, the prevailing idea was to build a way out of the problem. Indeed, these structural measures provided the foundations for a more sophisticated approach to flood damage reduction. The Mississippi River and Tributary Plan (MR&T), which was the USACE response to congressional authorizations of the 1930s to deal with devastating flooding, clearly acknowledged and included 'room for the river' as part of this highly successful 90 year program (Holmes, 1972; Arnold, 1988; Rosen & Reuss, 1988).

By the 1940s–1960s the view on flood management began to broaden. As early as the mid-1800s some realized the problem was settlement and cultivation in the floodplain. White (1945, 1969), along with Arthur Maass (1951, 1962) at Harvard and others developed alternative approaches to water resources and flood management. In 1953, the first major test of these land use and flood control measure ideas was undertaken by TVA. In 1954, the Watershed and Flood Prevention Act, of the SCS in the Department of Agriculture, was passed. In the 1950s and 1960s there was also a broader movement for more water resources coordination with a national water resources council (WRC) and a series of river basin organizations (White, 1969).

In the late 1960s and early 1970s, FPM began to take root as a complementary approach to structures as a way to manage the nation's flood problems. In 1968, Congress enacted the National Flood Insurance Act which established the National Flood Insurance Program (NFIP). The NFIP encouraged the widespread adoption of floodplain regulatory controls to discourage further development of damageable improvements in the nation's floodplains, and established a threshold of a 100-year flood zone. Supplementing structural approaches, FPM was meant to reverse pressures on development in floodplains.

Indeed, billions of dollars of flood damages were prevented and many lives saved. However, lacking a unifying policy, this period was marked by fragmentation of responsibility, differing and conflicting missions and roles, and even a certain amount of competition among agencies at the federal level. Other federal agencies became involved in urban flood management, including the recently created Housing and Urban Development Agency (HUD) in 1965. This led to further fragmentation of the federal roles and responsibilities.

During the 1960s, ideas about the best ways to reduce flood losses began to change. Continually rising flood losses in spite of a substantial investment in flood control, coupled with an emerging environmental ethic, brought shifts in thinking about floodplains. The US Congress directed the President to formulate a framework for managing the nation's floodplain areas. Numerous federal, state, local and private initiatives in the form of legislation, executive orders, interagency and intergovernmental agreements, comprehensive plans, public awareness campaigns, land trusts and scientific research all combined to form a broad foundation of principles for managing flood-prone areas.

In 1966, 'A UNP for Managing Flood Losses' was submitted to Congress (US Federal Task Force on Flood Control Policy, 1966). It recognized the need for a unified approach and for new procedures and made recommendations for initial federal actions, including legislation, specific studies, and new programs for collecting and disseminating flood related information.

In 1976 the US WRC produced the report, 'A UNP for FPM' (US Water Resources Council, 1976). The report described a framework for managing the nation's floodplains (not just the losses resulting from flooding), identifying strategies and tools for doing so, and made recommendations for recognition and acceptance of the program at all levels of government. Federal water resources planners were instructed to devise non-structural flood protection plans as viable alternatives to structural measures.

In 1979, a Federal Interagency Task Force on FPM updated and refined the UNP. This revision incorporated federal concerns with the natural and beneficial values of floodplains, responded to the President's policy, expanded the strategies, tools and conceptual framework accordingly, and emphasized the lack of adequate technical and procedural information to guide floodplain decision makers. The report was updated again in 1986 by an Interagency Task Force. It included the use of the federal interagency hazard mitigation teams, passage of the 1982 Coastal Barrier Resources Act restricting federal expenditures that might encourage development of coastal barriers, and completion of two major National Science Foundation studies on flood hazard mitigation. The report made explicit recommendations for federal support of state and local initiatives.

In 1994, a revised conceptual framework called for managing floodplains as integrated systems for both human activities and natural functions. It also listed specific national goals, set a timetable for their completion, and incorporated benchmarks by which progress can be measured. While all these federal task forces represented important conceptual milestones, the advocacy was largely in the federal domain, via Executive Orders. Congress did not follow up with meaningful comprehensive legislation to transform these conceptual ideas into a permanent footing, and local communities continued to favor structural solutions, which were seen to be more reliable and permanent. This created an uncomfortable dichotomy for the federal water management agencies – the local sponsors generally preferred structural solutions, while the federal entities promoted non-structural approaches. Meanwhile, Congress, which represented their constituents, kept funding studies and structural solutions (US Federal Interagency Management Task Force, 1994).

With these trends as a backdrop, the extreme floods on the Upper Mississippi River in 1993 spurred yet another major re-assessment of the nation's flood damage prevention and mitigation policies. The US President established an Interagency FPM Review Committee to delineate the major causes and consequences of the 1993 Midwest flooding, and to evaluate the performance of existing FPM and related watershed management programs. The Review Committee issued its report, 'Sharing the Challenge: FPM into the 21st Century', in June 1994. Later in 1994, the Federal Interagency FPM Task Force, established by the WRC, issued a revision to 'A UNP for FPM' (US Interagency Floodplain Management Review Committee, 1994). The newly revised UNP for FPM included four recommended national goals, each accompanied by several objectives.

However, most of the recommendations of 'sharing the challenges' were not completed, either by the federal agencies or by Congress. Indeed, the Nation's substantial residual flood problems are more difficult to solve than ever.

Today, over 20,000 localities regulate their flood-prone areas, making federally administered flood insurance available to their citizens. While most of these communities and extensive rural floodplain areas are subject to some frequency of flooding, only 10–15% of the structures in these areas are protected and few have a high degree of protection. Nevertheless, evidence indicates floodplain regulatory controls, as prompted by the NFIP, have deterred the development of damageable property in the nation's floodplains. However, such controls are not universal and not always strongly enforced. As a result, growth in damageable development continues, albeit at a reduced rate. Although these controls and the NFIP have been incentives to state and local governments to adopt regulatory controls, the flood insurance itself, along with the recommendations of the various committee reports, have not been a panacea (US Federal Interagency Management Task Force, 1994; ASFPM, 2004; Hessler, 2000; US FEMA, 2000).

3. Current situation: FPM, flood damage reduction policies and risk tolerance

3.1. Defining risk: multi-objective water resources management versus risk-based management

Today, DRR is underpinned in the USA by risk-based decision making. Risk-based decision making is seen as distinct from traditional water resources multi-objective decision making. The main difference in the two approaches appears to be the reality that the federally dominant 'top-down' comprehensive river basin oriented planning governance structures are dissipating and being replaced by more 'bottom-up' local, sub-state level governance decision making. This transforms the traditional water-based national economic development paradigm to one of sustainable development. This new paradigm maximizes social well-being, public safety and risk-reduction strategies for the local populace, subject to numerous environmental constraints (and preferred solutions) imposed by the regulatory agencies (Stakhiv, 2011).

In the past, growth was stimulated by federal economic and infrastructure investment policies, with generous federal subsidies. Today federal subsidies, regulatory and grant programs are used to transform development into what is known as 'green growth'. Whether this shift will result in more adaptable and resilient communities in the USA remains to be seen.

Risk-based decision making and management has also changed from a normative engineering-standards based approach, which relied substantially on flood event probabilities and engineering design infrastructure failure probabilities, to an 'analytical-deliberative informed consent' decision approach (Stern & Feinberg, 1996). In the past, flood risks were explicitly dealt with in terms of engineering-based threshold standards, such as the '100-year flood', or 'standard project flood' or 'probable maximum flood or hurricane'. The new 'risk-informed decision making' culture is far more complex. It is based on risk–cost comparisons and tradeoffs among various options. It engages the affected public.

This evaluation approach requires a great deal more complex and technically sophisticated information and the attention of the public – not just of analysts and decision makers. The public is also asked to bear more of the residual risks and costs, without often fully understanding the consequences.

DRR is not just an issue of economic planning objectives and engineering design standards, but one of evaluating the performance of risk reduction measures and benefits of numerous combinations of policies, regulations and structural solutions. Coupled with the large and growing uncertainties associated with climate variability, this analysis creates a cascade of growing uncertainties in decision making, that could be beyond the analytical capabilities of most policymakers, decision makers and most importantly, the public (Stakhiv, 2011). However, quantifying risk and uncertainty is at the center of DRR approaches to increasing resiliency. And herein lies the core dilemma. Engineering design standards already incorporate a socially determined 'acceptable' level of social risk tolerance, and they have the virtue of being uniformly applied. Risk-based decision making at the local level is neither replicable nor uniform. It is not at all clear whether a collection of loosely connected local solutions, with varying degrees of risk and uncertainty, can guarantee either robustness or resilience.

The National Research Council (NRC) of the US Academy of Sciences published a primer on 'Understanding Risk' (Stern & Feinberg, 1996). They understood that coping with risk situations is both complex and controversial, noting that numerous techniques have been developed to better characterize risks in order to make informed decisions about hazards to human health, welfare, safety and the environment. Some argue that as the literature and methodologies for risk analysis expanded during the past two decades, the quality of decisions suffered because of the technical complexities associated with such analysis. There was no easy way to transform complex analytical characterizations of uncertainty into forms needed for 'public decision making' (Stern & Feinberg, 1996).

The NRC report offered several principles for undertaking risk analyses:

- Risk characterization should be a 'decision-driven activity', directed toward informing choices and solving problems.
- Coping with a risk situation requires a 'broad understanding' of the relevant losses, harms, or consequences to the interested parties.
- Risk characterization is the outcome of an 'analytic-deliberative process'.
- Success depends on deliberations that formulate the decision problem, guide the analysis to improve decision participants' understanding, seek the meaning of analytic findings and uncertainties, and improve the ability of interested and affected parties to participate effectively in the risk decision process.
- The analytic-deliberative process should be 'mutual and recursive'.
- Each organization responsible for making risk decisions should work to 'build organizational capability' to conform to the principles of sound risk characterization (Stern & Feinberg, 1996).

Historically, the US federal water agencies, such as the Corps and Bureau of Reclamation, have adhered fairly closely to these principles via the former WRC Principles and Guidelines (P&G) issued in 1983. Risk and uncertainty have always been major components of the decision-making process within agency procedures. They are inherently part of a broader analysis of water-related objectives and means. The difference in the new NRC approach, versus that of the broader and more encompassing WRC P&G approach, is that the NRC approach views problems through a risk and uncertainty prism, much the same as NEPA views problems through a single-objective environmental prism.

One important component of DRR places the onus of understanding the complexities of risk reduction options on the public and local officials. In order for the new paradigm to succeed, the USA, like other nations, needs to dramatically improve risk communication policies and procedures for DRR. Some argue that the 100-year flood should be termed the 1% or high risk flood, and the 500-year event becomes the 0.2% or extreme risk flood. Some believe, however, that the movement to a risk-based water resources planning and decision-making framework and away from designing to pre-determined engineering design standards may result in more structures being built in flood hazard zones, increasing exposure and susceptibility to flooding.

As Figure 2 shows, structures at 1% risk of flooding actually have a 25% chance of being flooded during the typical 30-year life of a home loan (mortgage). There is a 1% chance that the same structure will have a fire – yet almost all homes and businesses have fire insurance, and less than 25% have flood insurance. A 'residual risk' will always exist, even with flood control structures and flood insurance, for there will always be events larger than the 'acceptable' design storm event – i.e. the inevitable 'acts of God'.

What happens if the populace decides to accept a 'tolerable degree' of risk that is greater than engineering design standards, based on their calculation of a risk–cost optimum? In a democratic society, the key is to link risk with responsible behavior; to encourage the active choice and acknowledgement of flood risks, versus a passive reliance on institutional actions. However, the public must somehow be fully aware of both the risks and consequences. Defining 'tolerable risk' and 'residual risk' no longer remains a scientific or technical exercise, as it quickly moves into the realm of political choices, where aspects of equity and social justice are joined with a myriad of other aspects of ethics and morality. With climate change and adaptation, this becomes exceedingly more difficult because of increased uncertainties that complicate rational decision making.

Figures 3 and 4 contrast traditional and contemporary paradigms or views of risk management used in flood DRR. Today, water-related risk management starts with various stakeholders and behavioral and policy measures to build cooperative approaches to risk management. While residual risk has always been a reality, it is far more explicitly dealt with in contemporary management approaches. Most of the 'residual risk', i.e. that risk which cannot be ameliorated through structures, zoning policies or insurance, is largely borne by individuals, homeowners and small businesses. Yet the degree of 'residual risk' is rarely quantified or explained to the public in terms they can understand. Residual risk is almost always underestimated because it is difficult to quantify a cascading series of highly interdependent measures, each of which has its own reliability characteristics and risk of failure. Mostly, though, a flood protection system is a collection of fragmented measures, including new building codes, zoning ordinances and structural measures that are implemented over a long period of time and loosely coordinated by multiple authorities. This comprises the definition of a 'brittle system'.

Risk reduction is clearly seen as collaboration among levels of government and affected citizens. Stakeholders must become active participants in the process of defining, sharing, accepting and

Return Period (years)	Chance of flooding in any given year	Percent chance of flooding during 30-year mortgage
10	10 in 100 (10%)	96%
50	2 in 100 (2%)	46%
100	1 in 100 (1%)	26%
500	0.2 in 100 (0.2%)	6%

Fig. 2. Characterizing flood risk (USACE, 2014).

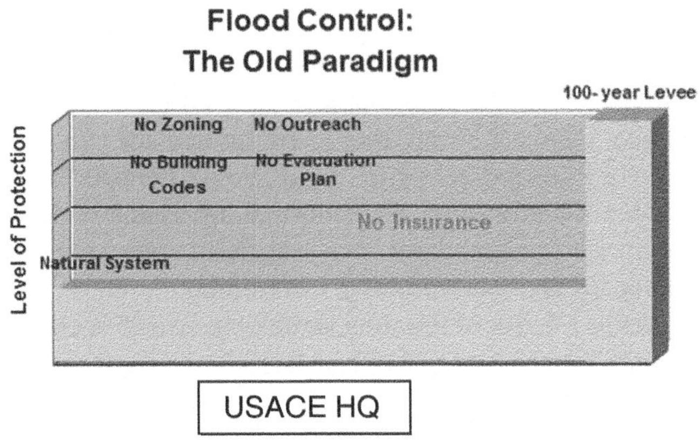

Fig. 3. Flood control: the old paradigm.

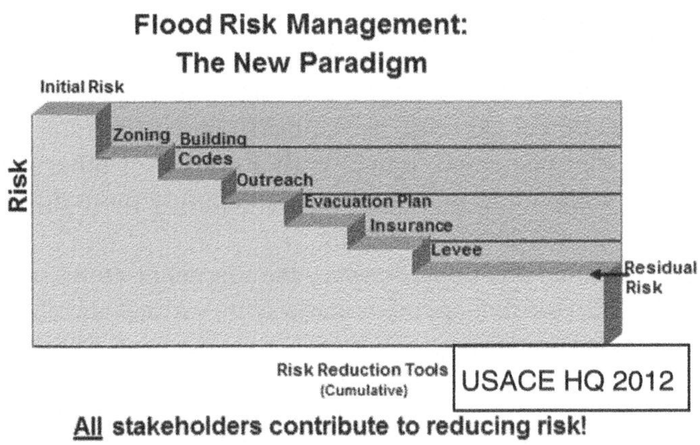

Fig. 4. Flood risk management: the new paradigm.

choosing levels of risk. In this sense, the public must shift from the traditional view of waiting for the engineers or experts to 'solve the problem', or the illusion that somehow structures will eliminate risks. Since standards of risk are very much based on consensus, how we make decisions about risk management is central to the health of democratic systems.

3.2. Flood management and socio/economic development

Since the Flood Control Act of 1936, the nation, through the US Army Corps of Engineers, as the principal US national flood management agency, has invested about $100 billion (in 1996 dollars) to construct, operate and maintain flood control structures. These include approximately 400 major lake and reservoir projects, over 16,000 miles (25,750 km) of levees and dikes, 400 miles (644 km) of coastal shore protection and thousands of smaller local flood protection projects. Corps projects, alone, have prevented nearly $600 billion in flood damages since 1950.

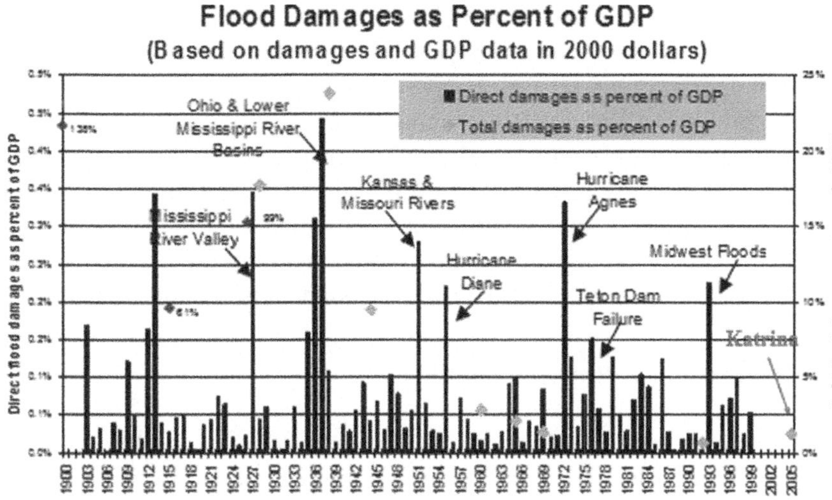

Fig. 5. Damages from water-related disasters as % of GDP[1]. Constructed from Mohleji & Pielke (2008), Pielke *et al.* (2008) and Shabman (1988).

Projects of other federal and local agencies also have prevented substantial, though lesser, damages. In addition to damages prevented, damages avoided as a result of regulatory controls and related incentives are presumed to be substantial. Although no one authoritative assessment of damages avoided exists, one study of small to medium size US cities found that floodplain land use controls substantially reduced the potential damages of foregone development, which might have occurred in the absence of the controls (ASFPM, 2004). Annual damages, while widely variable from year to year, have remained relatively constant, if not slightly decreasing, as a percentage of Gross National Product from 1903 through to 2005 (Figure 5).

Despite these efforts, there remains a substantial residual flood damage problem in the USA even though existing programs seem to be damping that problem. Emergency disaster relief costs are still high, averaging around $3 billion per year (1991–1997), and uninsured losses continue to mount. There are two aspects of the residual problem which are critical. One is the extensive unprotected development still remaining within the 100-year floodplains along the nation's streams, despite all flood control, FPM and regulatory efforts. The other is the continuing development just outside of the 100-year floodplain, which is not subject to floodplain regulations, but is subject to less frequent (for the particular stream and location) but more damaging extreme floods (Weiner, 1996).

The relative wealth of each household and business has been increasing dramatically, so that for the same magnitude flood event of 50 years ago, compared to today, the damages would be expected to

---

[1] Flood damages as % of GDP (in 2000 dollars) = [unadjusted flood damages data (in current dollars)/deflator index] / GDP (in 2000 dollars). Sources of unadjusted flood damage data (in current dollars): direct unadjusted flood damages: NOAA. Data are for water years, from 1 October to 30 September; total unadjusted flood damages data: Burton & Hicks (2005) paper, adjusted from data provided in Pielke & Landsea (1998). Source of estimated deflator indices (to convert current dollars into 2000 dollars): from 1929 to date: http://www.flooddamagedata.org/national.html; prior to 1929: assume 2% increase (similar to the assumption used in Pielke & Landsea (1998)). Source of estimated GDP data: EH.NET 2005.

increase geometrically, rather than in line with simple inflation. Also, development in upstream flood 'free' areas and channelization of streams continues to contribute to increasing storm water runoff rates thus increasing flood peak potential beyond previous expectations.

The US experiences show that flood DRR measures are closely tied to socio/economic development and social stability. Damages, as a percentage of gross domestic product (GDP), have gone down dramatically. This indicates that in many cases the investment in DRR methods provides enough resilience to reduce the potential of political–social failure under the stress of water related disasters. However, DRR investment policies must also guard against 'inducing benefits' by creating socio-economic 'moral hazards' activity in high risk areas (US Office of Management and Budget, 1998).

3.3. The National Flood Insurance Program (NFIP)

The NFIP is now the cornerstone of US water DRR Policy. Some of the key rules for NFIP include: no development in a defined floodway; no residential living area below 1% flood level; no non-residential development subject to damage by 1% flood; no rebuilding of flood-damaged property if an event is below 1% flood probability and if damage is 50% or more of structure value.

Community measures must meet minimum Federal Emergency Management Administration (FEMA) standards and include: zoning, subdivisions, building requirements, special purpose ordinances, outreach, education, and others.

3.4. Barriers to balancing structural and non-structural measures

There are many barriers to adopting non-structural approaches and balancing them with structural approaches, apart from the difficult task of quantifying risks and uncertainties. There is often no full or even adequate accounting of the environmental quality and non-monetary social benefits of non-structural measures. The flood risk reduction values of natural systems are just now beginning to be described and evaluated as new analytical techniques and data improve. Existing procedures used for benefit–cost analyses are often based on certain assumptions that limit non-structural benefits. Emergency flood relief and recovery payments by the Federal Government (FEMA) create incentives against non-structural measures. FPM is not being pursued in a comprehensive fashion at all levels of government. Those who live and work in flood zones are not paying the proportional share of those location decisions as a result of less than actuarial flood insurance premium pricing. The national policies on disaster response and assistance are not aligned to flood management principles – and are actually still creating incentives for locating in floodplains (US Water Resources Council, 1983; US CEQ, 2014; Holliday *et al.*, 1998).

3.5. Overcoming barriers – the debate on policy directions in the USA

To overcome barriers to a sustainable flood policy, the USA is moving toward using 'likely future conditions' to make risk analysis more realistic. Federal, state and local authorities are trying to include disaster resiliency in community planning. Such resilience should be built on flood control measures, flood resistant construction, storm water management, community zoning subdivision and regulations (Holliday *et al.*, 1998).

To increase the use of non-structural measures, the USA is advocating buyouts (such as in the post 1993 flooding), elevating buildings, buffer zones and levee setbacks. Some advocate keeping vacated land in public ownership. Everywhere, natural storage capacity is being sought. Building codes are being redefined for structural wind damage and elevation requirements. New construction in flood risk areas should include 1–3 feet (0.3–0.9 m) of freeboard above base flood elevation. Some advocate the use of confidence levels (90–95%) for flood peak flows predictions.

Many barriers to reform exist in agricultural policies. Crop losses often exceed urban losses and may comprise over half of flooding damages borne by taxpayers. The USA also needs to examine the prevalence of repeat areas of losses. Many advocate instituting voluntary permanent easements programs. Policy makers should perhaps deny subsidies and disaster payments if refused permanent easement purchases. More buffer zones are needed.

There are major barriers to reform in US coastal policies, including existing policies that often foster rather than discourage construction on coasts. The NFIP has not been particularly effective in coastal areas, but there is a need to integrate coastal areas into the NFIP and place surcharges on areas subject to erosion and establish setback requirements. To sustain existing infrastructure, shoreline erosion responses need to shift from jetties and sea walls to beach nourishment, although in densely populated urban areas, there may be no viable alternatives to structural measures. However, these are expensive solutions that need to be coupled with revised cost sharing formulas between the federal and state governments, tied to setbacks, acquisition strategies and increased public access to improved beaches. Also, such policies require land use regulations which are in the legal purview of sovereign states and local governments.

Barriers also exist with current data and technology. There is no single entity with responsibility for collecting and storing data about flood frequency and magnitude, or recording damages. The number and value of structures in flood hazard zones are not known, but new geographic information system (GIS)-based systems that integrate various databases in 'real time' with operations models are significantly overcoming this barrier. The USA needs better data on repetitive losses to structures. FEMA is improving methods for estimating flood damages but, at the same time, the stream gaging network is shrinking. New modeling that includes unsteady flow conditions, levee breaches, split flows and unstable land forms, and debris flows are being developed. However, design manuals need to be updated to include alternatives to structural methods and newly developed bioengineering approaches.

The flood insurance program and flood policy in general needs to deal with repetitive loss problems. Two percent of policies under NFIP accounted for 32% of losses and received 38% of pay outs from the National Insurance fund. Three hundred high loss repetitive communities, in 35 states, or 1.6% of the 18,700 communities in the NFIP, have 31,574 structures that are less than 1% of the total insured but receive 29% of all NFIP loss payments. There is a need to move to actuarially based premiums or to adjust according to use of mitigation. Replacement costs are better for coastal situations and market value for riverine situations.

4. Three recent cases of water related disasters: 2005 Katrina storm surge, 2011 record Mississippi river floods and 2012 Superstorm Sandy storm surge

4.1. Case I: Hurricane Katrina 2005

In August 2005, the people on the gulf coast of the United States suffered a great disaster which the world watched. During Hurricane Katrina, 1.2 million people were evacuated temporarily and 1.5

million displaced for an extended period of time. Nearly 1,500 people died during the event. Today about 70% of the population of New Orleans has returned but the percentage returning to the hardest hit and poorest areas is much lower.

Katrina was one of the greatest natural disasters the USA has ever suffered. The storm disrupted the social fabric of the entire region: employment, trade, travel, education, health care, worship and community. Quantifiable consequences include direct property losses of over $16 billion in residential losses; $2.4 billion in commercial losses; and $4.4 to $5.6 billion in infrastructure losses, excluding the hurricane protection system (HPS). Total losses were more than triple those from any previous disaster in the New Orleans area. Recovery, measured by the return of the population and resumption of business activities, was hampered by damage to infrastructure and affiliated public welfare and services. According to the Interagency Performance Evaluation Task Force (IPET) the mass exodus of the region's inhabitants, combined with structural devastation and the demands of rebuilding, has fundamentally changed all aspects of the population's social structure (USACE, Institute for Water Resources, 2008). About 50% of property owners suffering flood damage were insured, which is higher than the average in the USA.

Katrina overwhelmed most of the HPS and significantly exceeded the design assumptions. The system suffered 50 breaches, causing approximately two-thirds of Katrina-related flooding. The remaining one-third was caused by overtopping and rainfall. Forty-six breaches were caused by water levels overtopping (exceeding the elevations of) floodwalls or levees. IPET noted that four I-wall segments failed below design levels along the 17th Street canal because of unanticipated failure modes (Figure 6).

It could have been much worse. If the record Mississippi flood of August, 1993, had coincided with a hurricane of Katrina-type magnitude, it would have been comparable to the great flood in Bangladesh in 1988, when monsoon flooding coincided with a cyclone (hurricane) in the Bay of Bengal. More than 30 million people were homeless, over 100,000 died and nearly 70% of Bangladesh was flooded. Louisiana has a similar topography and situation, and Dhaka, Bangladesh, has comparable physiographic characteristics to New Orleans.

As mentioned above, federal commitment to flood protection increased along the Mississippi and around New Orleans steadily during the last century. In fact there were large storm surges in 1812,

Fig. 6. Category 5 hurricane.

1915, 1947 and 1965. However, it was not until 1955, after Hurricane Elena struck New Orleans, that the US Congress authorized \$85 M for storm surge and hurricane protection with an estimated completion by 1979 (Figures 7 and 8). New Orleans thus started two levee systems: one for spring floods and one for fall storm surge events. The new storm surge or Hurricane Protection project was designed for a category 2+ event (Simpson-Saffir scale) with a statistical return period of 200–300 years. Two alternatives designs were considered. First a Barrier Plan with two rings of protection, consisting of an inner ring of 9–13 foot (2.74–3.96 m) levees with an outer storm surge barrier with flood gate. The second alternative

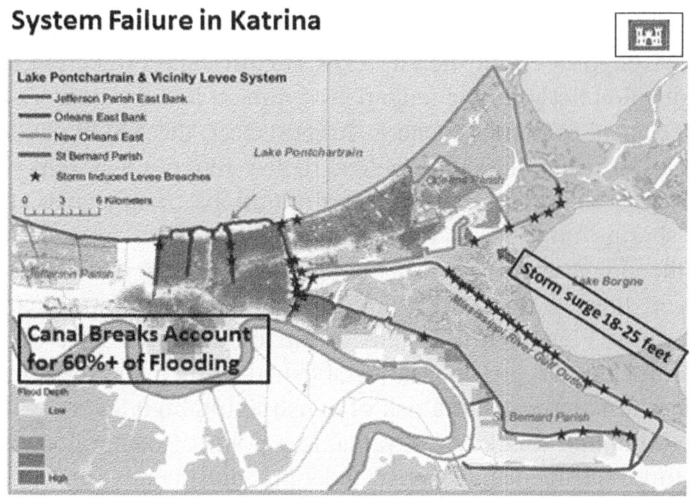

Fig. 7. System failure in Katrina.

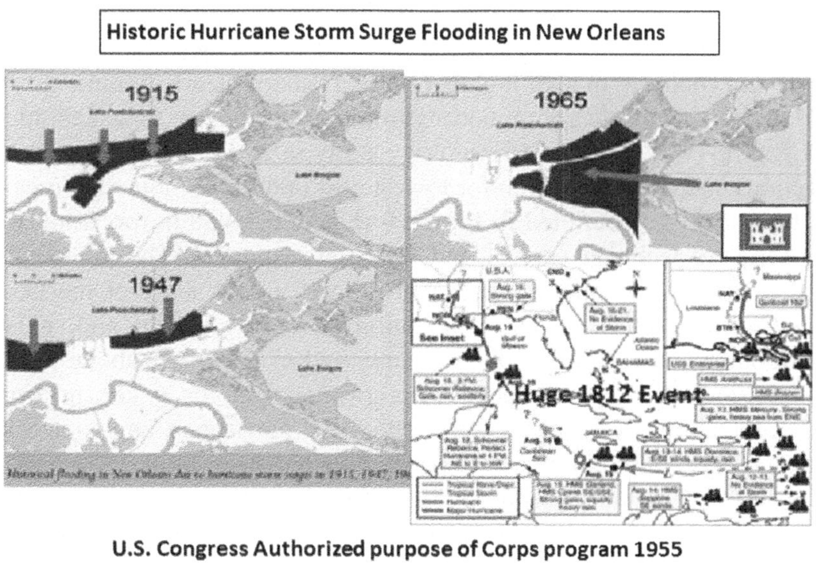

Fig. 8. Hurricane storm surge flooding in New Orleans.

was a High Levee Plan which consisted of two inner rings of high levees at 16–18.5 feet (4.88–5.64 m). (US IPET, 2009; USACE Institute for Water Resources, 2008).

In 1975 and 1977 public opposition and environmental lawsuits over the two-ring plan emerged and it was abandoned. By 1982, the costs rose from a projected $85 to $800 M with an estimated 2015 completion. From 1995 to 2005, federal funding declined from $15–20 to $5–7 M/year.

Building in the Mississippi Delta had been increasing with the construction of extensive levee systems. There were also declining sediment loads in the river and canals were being dredged for navigation or in support of mineral extraction. The combination of these actions led to considerable land subsidence in New Orleans, which was already lower than sea level. New Orleans was essentially an enclosed basin or bowl that was slowly subsiding even further below sea level (ASCE, 2007).

In 1955, when the US Congress established a USACE Hurricane Protection mission, a 70–30% cost sharing for hurricane protection between the federal government and the state was decided. The resulting HPS was constructed piecemeal over four decades. In retrospect, the ASCE studies noted that this led to a failure to see that the sum of many parts did not lead to a system nor a system-wide approach to design or operation. The result was changing technical configurations, rising costs, project extensions, and unclear mixing of cost sharing interests and technical considerations between state, local and federal entities. For example, the potential of a storm surge forcing water back through the canal outfalls was recognized. However, there was a long standing culture of pumping water out of New Orleans and fear of blocking outfalls. Plus such a system would incur higher costs to local interests. Thus a plan of high levees around outfall canals was established, which required far more federal money outlay. Engineering design tests indicated the risk of constructing high levees versus some outfall pumping was about the same, although this key test remains a point of contention for some.

During this extended time, subsidence and vertical datum adjustments were not considered. The American Society of Civil Engineers (ASCE) noted that the Standard Project Hurricane definition that was developed after the 1965 hurricane was never updated and design criteria were based on assumptions and conditions at the beginning of the HPS implementation with no systematic updates being made.

It appears that neither system-wide considerations of residual risks or deliberate treatment of system reliability issues were considered. Also, there was a lack of adaptive treatment of risk and reliability over the many iterative design changes. In addition, there was a lack of clearly communicated residual risk or of the monumental scale of possible consequences from system failure to stakeholders and the public. According to the USACE-IWR Decision Chronology study in 2009, this incremental decision making, federal-state interactions and constant search for lowest cost options combined to create a hurricane protection policy 'perfect storm', as outlined in Figure 9 (USACE Institute for Water Resources, 2008).

Katrina led to a series of technical, policy and financial responses. The immediate recovery period was followed by new flood maps, the rebuilding of homes, a new study of the Louisiana Coast (the USACE Louisiana Coastal Protection and Restoration (LACPR) Study) and, most importantly the new paradigm for flood risk management policy mentioned in Section 1 above.

Technical responses included repair of levees/floodwalls, pumping water, restoration of navigable waterways, upgrade protections to standards for category 3+ hurricanes, and an accelerated completion of protection measures. Also, a massive new surge barrier has been built.

Policy, institutional and regulatory responses included all levels of government and public participation in the IPET, the LACPR study and local governance Levee Board Reform in New Orleans. The IPET (US IPET, 2009) conducted a forensic analysis of what happened. The IPET consisted of teams for: engineering and operation risk and reliability analysis; consequence analysis of Hurricane

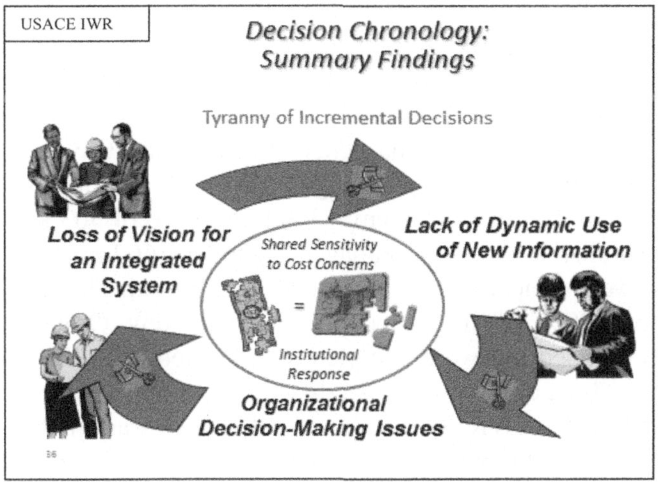

Fig. 9. Decision chronology summary findings.

Katrina; pumping station performance; analysis of floodwall and levee performance; geodetic vertical survey assessment; storm surge and wave physical model – centrifuge modeling; storm surge and wave physical model – hydrodynamic forces; numerical model of Hurricane Katrina surge and wave environment; interior drainage numerical models; and data collection and management – perishable, systems data, info management. As Figure 10 below shows, the process included considerable peer review and experts outside the immediate agencies involved (US IPET, 2009).

The IPET concluded that as Katrina progressed slowly across the Gulf of Mexico, the storm gathered enormous quantities of water. A Category 5 hurricane with 175 mph (282 km/h) winds located 170 miles (274 km) from landfall, Katrina dropped to a Category 3 hurricane, with wind speeds of 127 mph (204 km/h), when it came ashore at Buras. But the storm brought with it record levels of

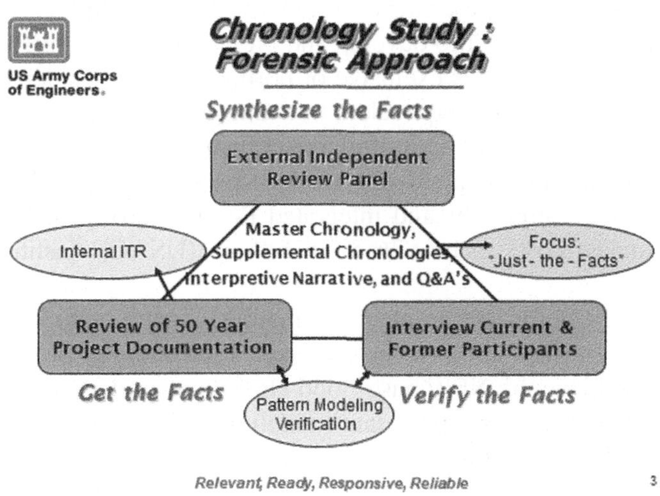

Fig. 10. Chronology study: forensic approach.

surge, waves and rainfall, of magnitudes for hydrodynamic aspects of storm strength not factored in the Saffir-Simpson hurricane scale (Category 1–5) (US IPET, 2009).

The IPET noted that although Katrina, with wind speeds of 127 mph, was classified as a Category 3 hurricane at the time of landfall, its levels of surge and waves exceeded those of any previous storm striking the North American continent. Katrina generated surge and waves causing major breaches in the HPS (described above) and significantly damaged 34 of 71 pumping stations designed to move water out of the city, as well as 168 of the system's 350 miles (270 of 563 km) of protective structures, such as levees and floodwalls. It severely damaged 41 miles (66 km) of protective structures and brought record rainfall of over 14 inches (0.36 m) in a 24-hour period. The rainfall, combined with overtopping, contributed about 30% of the floodwater in the New Orleans area. Sixty-five of 73 neighborhoods in the city flooded; 34 were completely inundated (US IPET, 2009).

In addition to forensic study, a chronological study of all decisions made from the mid-1960s to the event was prepared by the USACE-IWR. This study revealed there was an absence of dynamic practices and processes to ensure that new knowledge and technologies were incorporated into project design; updated considerations of project reliability and sustainability under extreme events was not manifested in project design or system performance evaluations (USACE Institute for Water Resources, 2008); over time there had been a loss of an integrated system perspective as the original unified vision devolved into a series of incremental segments over a protracted design and construction period; key design and performance decisions were made at field level, with limited oversight, and made with the presumption these were purely technical decisions, and without evidence of USACE-HQ awareness of policy implications; and the explicit consideration and open communication of system-wide residual risk, consequences and reliability were not evidenced in record.

The Decision Chronology Study also revealed that the US federalist system produced serious unintended consequences of incremental policy decisions over 20 years. For example, cost sharing requirements for the project, combined with local and federal budget constraints, motivated partners to constantly seek ways to reduce their cost burden. Cost concerns blurred distinctions between technical, policy and fiscal considerations. Technical decisions were significantly influenced by local water management objectives. Rather than staying within the defined system design, this incremental decision making created a piecemeal storm surge protection 'system' in name only.

The post-Katrina LACPR study sought to develop and present a full range of flood control, coastal restoration and hurricane protection measures exclusive of normal policy considerations for south Louisiana and consider protection for a storm surge equivalent to a Category 5 hurricane. The study acknowledges different levels of protection and multiple lines of defense (Figure 11) to include non-structural measures. Most importantly, it put forward a conceptual vision for a sustainable coast that emphasized sustainable ecosystems, flood protection, communities and integrated flood control projects, and coastal restoration initiatives. The study estimated costs to be $76–$136 billion. (USACE Institute for Water Resources, 2008).

After Katrina, the US Army Corps of Engineers operated a recovery activity called Task Force Hope. This included modifying authorized projects in southeast Louisiana to provide hurricane, storm and flood damage reduction to achieve the NFIP base flood elevations, under an updated definition of a Category 3 hurricane as based on the most up-to-date modeling and risk-based designs. The rehabilitated system was cost shared at 65% federal and 35% non-federal contributions and completed in 2011.

Financial responses included $2.28 billion in Congressional appropriations to the Corps for projects in Louisiana, Mississippi and Texas. Of this, $1.4 billion went to New Orleans. FEMA administered a

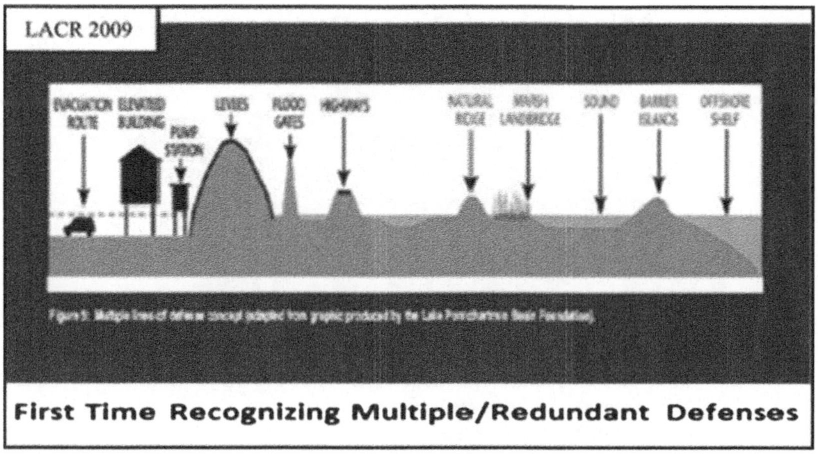

Fig. 11. Multiple/redundant defenses.

mitigation grant program with reimbursements of up to $100,000 to raise the lowest floor of a residence to the FEMA base flood (100-year) elevation. Eligible homeowners affected by hurricanes could receive up to $150,000 in compensation for losses to get them back into homes. The federal recovery program funded 126,780 applicants and paid out $8.33 billion.

While new methods of public communication and humanitarian assistance with housing began, many policy questions remained. For example, there is still no significant discussion of the overriding social issue of whether people should be living in high risk, flood prone areas or whether people should move back – if so where or how? What is the public interest for long-term planning of DRR infrastructure in this complicated federal system and how should it be factored into decisions? Does the influence reside with those who pay, such as sponsors of projects only, or with the broader public who subsidize risky behavior through taxes? How can regional decision-making approaches to recovery that combine revenue generated from a region with cross subsidies be realized? Are there special rules for at-risk urban areas? What is the best form of risk insurance? Given the increased vulnerability of coastal cities, is it possible to consciously relocate cities? Who do we protect – those whose properties are most highly valued, according to existing benefit–cost procedures? How do we protect those without property or money for insurance? There is a need to view the integration of relief, response and prevention as an holistic risk management endeavor, where citizens actively participate in choosing levels of acceptable risk. But this has proven tricky because 'acceptable risk' is difficult to understand, equivalent to choosing among alternative modes of cancer treatment – each with different probabilities of success and numerous uncertainties.

4.2. Case II: 2011 Mississippi flood

In late summer of 2011, the Mississippi River reached some of the highest recorded levels in US history. USACE managed the flood, with other federal and state local agencies, through the MR&T project which had been constructed over the last 70 years and was about 90% complete at the time of the 2011 floods (USACE, MVD, 2012) (Figure 12).

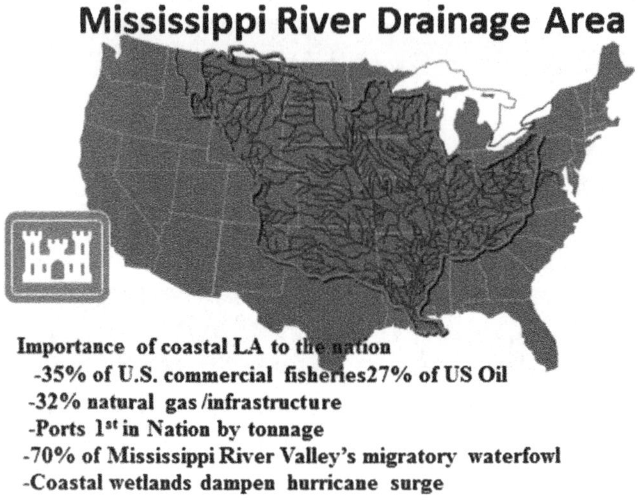

Fig. 12. 2011 Mississippi flood.

The MR&T was authorized and designed after the great 1927 flood. In the 1927 event: 500 people died; 600 thousand people were left homeless and went to refugee shelters; 41,000 buildings were destroyed; rail lines were cut throughout the center of the USA; waterways transport of key commodities were disrupted for months; and east-west telecommunications and postal services were put into chaos. As a result, one of the most important realignments in political demographics in US history occurred.

During the 2011 flood, the MR&T, designed in the 1920s, combined 'room for the river', structural and non-structural approaches, and essentially worked as planned and designed. (USACE, MVD, 2008, 2014). It confined flood events to natural and selected agricultural land overflow areas, protected more than 10 million acres of land and one million structures, and prevented over $110 billion in damages. The purposes of the MR&T included navigation, flood control and promotion of commerce. It is one of the largest flood risk reduction projects in the world and has cost $13.9 billion covering 36,000 square miles (93,240 km^2) of floodplain. It includes levees, floodways for excess flows, channel improvements, tributary bank stabilization, dams and channels. These features were designed to manage a 'standard project flood' that would be 11% greater than the 1927 flood. Between 1928 and 2004 $425.5 billion was invested (USACE, MVD, 2008, 2014) (Figure 13).

The 2011 event was close to the size of the historic 1927 event. By contrast, in the 2011 event over 4 million people were protected. It realized $478.3 billion in prevented flood damages, which resulted in a 34:1 return on public investment. The 2011 event set river stage and flow records over much of the lower Mississippi River basin. During the flood, USACE operated its floodways and backwater overflow areas, which caused 6.35 million acres to be flooded, though 1.5 million acres of floodways and backwater areas were not used (USACE, MVD, 2008, 2014).

This was the first time all floodway components of the MR&T system were operational at the same time. Opening the floodways reduced the river's crest by 2.5 feet (0.76 m) between Baton Rouge and New Orleans and protected a 200-mile (322 km) corridor of the nation's business resources, including the cities of Baton Rouge and New Orleans. The MR&T system performed as designed and reduced the risks of flooding for more than 4 million people (USACE, MVD, 2008, 2014).

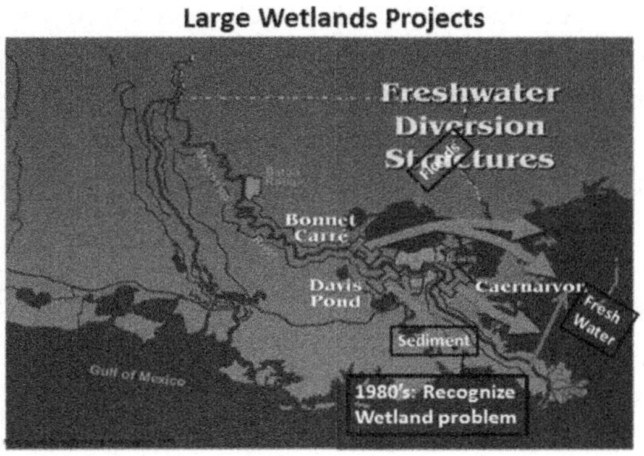

Fig. 13. Large wetland projects.

4.3 Superstorm Sandy

In October 2012, people up and down the eastern coast of the United States suffered enormously from Superstorm Sandy storm surges that flooded major urban areas and severely damaged hundreds of miles of shoreline. Damages resulted from heavy rainfall as far inland as the Midwest, storm surges exceeding 13 feet (3.96 m), and wind gusts of 80–90 mph (129–145 km/h). One hundred and seventeen people in the USA were killed and 650,000 homes were damaged or destroyed. Damage estimates for the New Jersey (NJ) – New York City (NYC) area alone exceeded $60 billion. Political arguments of who will pay and how much should be rebuilt continue (Figure 14).

Tunnels into NYC were flooded, effectively cutting off the island of Manhattan. Hundreds of thousands of people lost power, many for months. At the peak, USACE controlled 162 pumps un-watering over 475.5 M gallons of water. This included the Brooklyn Battery Tunnel, the longest underwater vehicular tunnel in North America and second longest in the world. In addition, USACE supported efforts in NY

Fig. 14. Effects of Hurricane Sandy.

and NJ to establish emergency power in strategic sites such as hospitals, Wall Street financial district, and essential services. USACE also removed 675,692 cubic yards (516,603 m^3) of debris (as of 25 Feb 2013). USACE employed almost 4,000 people in these efforts while expending over $250 million.

Post-Sandy, the federal government established the joint National Oceanic and Atmospheric Administration (NOAA)-USACE Sandy Infrastructure Systems Rebuilding Principles (US NOAA & USACE, 2013):

1. work together to develop long-term strategies,
2. improve coastal resilience,
3. increase awareness of risks and consequences.

The principles built on the value of integrated approaches to risk reduction and on incorporating natural and nature-based features, in addition to non-structural and structural measures. This followed the historical path of the Jadwin Report after the flood of 1927, which required 'room for the river' so that floodplains are able to carry flows greater than normal, and the Mississippi Coastal Improvement Project that followed Hurricane Katrina.

The first principle is to work together and develop long-term strategies. It calls for: involvement by stakeholders in scoping, development, implementation and monitoring of integrated solutions; working across the federal government to promote alignment of agency actions; leveraging partnerships to maximize all appropriate sources of funding, resources and expertize; working with state and local partners to sequence and focus rebuilding and restoration; aligning and delivering data, tools and information (e.g. physical, ecological, economic, etc.) in easily accessible formats; ensuring the federal government provides useful and timely technical assistance and information; and learning from the experience of Superstorm Sandy by conducting assessments of what worked and what did not.

The second principle is to improve coastal resilience. This principle promotes integration of natural and built systems: using adaptive management to encourage flexible decision making; seeking to modify critical infrastructure in response to changing conditions; advancing the understanding of a system-based approach and the benefits of the natural environment and its 'services', including coastal buffering; supporting sustainable economic activities and strengthening of existing social institutions; identifying economically-viable solutions that minimize impacts to the natural functions of floodplains and coastal ecosystems; working together to identify priority actions on a system or sub-system basis; and aligning investments to ensure water dependent uses of the coast, particularly ports and related infrastructure, supporting working waterfronts that are more efficient, safe, secure, resilient and environmentally sustainable.

The third principle is to increase awareness of risks and consequences. This requires: improving understanding among decision makers; encouraging risk-informed decisions that consider uncertain changes in the natural and built environments, including the effects of climate change, land-use change and coastal development; investment in risk communication efforts, including disclosure of risks that have not or cannot be mitigated in an economically feasible manner; encouraging coordination of common approaches to the characterization of risks, vulnerabilities and mitigation strategies; and developing and sharing user-friendly information and tools for assessing impacts, managing risks and risk/reward tradeoffs related to different project options.

Nevertheless, even though there was an interagency task force that attempted to coordinate the various ideas and plans of the federal, state and municipal agencies, each institution developed its own approaches to their respective responsibilities, which were difficult to integrate effectively into a comprehensive plan. Furthermore, the costs of the various solutions were well beyond the budget laid out by Congress. Very little has been implemented so far, with the Corps of Engineers shore protection component as one exception.

Sandy alerted the USA to the growing challenges of climate variability and urban design.

Indeed, the third National Climate Assessment (NCA) for the USA, released in May 2014, reports that climate change and its impacts are major threats to urban residents. It notes essential infrastructure systems such as water, energy supply and transportation will increasingly be compromised by interrelated climate change impacts. In urban settings, climate-related disruptions of services in one infrastructure system will almost always result in disruptions in one or more other infrastructure systems, and the vulnerability and adaptive capacity of urban residents will be threatened by pronounced social inequalities, such as differences in age, ethnicity, gender, income, health and disability (US National Climate Assessment, 2014).

The NCA's bottom line is that the nation's economy, security and culture all depend on the resilience of urban infrastructure systems. The challenge we now face, in NYC and other coastal cities, is how these cities can evolve to prepare for this level of inundation, before it happens. City government agencies and organizations have started adaptation plans that focus on infrastructure systems and public health. To be successful, these adaptation efforts require cooperative private sector and governmental activities, but institutions face many barriers to implementing coordinated efforts (US National Climate Assessment, 2014).

A recent independent analysis by the Rand Corporation (Finucane *et al.*, 2014) of how the federal agencies were implementing the Obama Administration's infrastructure resilience guidelines in reconstructing the areas damaged by Superstorm Sandy, showed expected disparities and confusion among the agencies. The seven principles in the guidelines are: (1) comprehensive analysis, (2) transparent and inclusive decision processes, (3) regional resilience, (4) long-term efficacy and fiscal sustainability, (5) environmentally sustainable and innovative solutions, (6) targeted financial incentives, and (7) adherence to resilience performance standards.

> *'Building community resilience via a 'holistic systems approach' (i.e. one that underscores the dynamic links among human, social, physical, economic and natural resources) was a new concept for some agencies and their grantees. For others, the focus on green infrastructure and nature-based solutions to deal with flooding and other storm impacts was a new pursuit'* (Finucane *et al.*, 2014).

The main challenges identified by the federal and state participants underscored common difficulties in complex interagency initiatives. For instance, finding the right level of specificity in the guidelines was difficult because they were devised to be applied broadly, yet were meant to guide specific actions across diverse sectors and locations. Diverse contexts also resulted in diverse interpretations and metrics for resilience, which made it difficult to know whether the guidelines were being followed and whether the intended progress was being made. Most federal participants expressed a strong desire for a more streamlined approach to prioritizing the myriad guidance, executive orders, frameworks and plans related to resilience. In other words, there was too much generic guidance, which hampered investment decisions.

More specifically, the following barriers to implementation were commonly voiced:

- Adjacent communities would opt for structural and non-structural solutions, compromising the robustness and reliability of the overall plan.
- There were many diverse interpretations of resilience, which hampered the search for viable solutions. Resilience could mean anything from readiness, preparedness, speed of recovery, to economic, social

or environmental resilience. Hence, there was no meaningful way of measuring system performance, after all desired solutions were implemented.

- Interpreting technical information and risk evaluation procedures was a distinct barrier. Federal regulatory agencies were particularly critical of certain solutions and the rationale used to justify certain measures.
- Numerous regulatory requirements and criteria varied considerably, and were not aligned with the resilience principles. The same problem existed with a range of grant programs (e.g. Small Business Bureau, HUD Community Block Development Grants, etc.).

5. Overall lessons learned from the three recent mega-disasters

(a) DRR comprises a special category of flood management, as it connotes extremely large events, with catastrophic human consequences and national level economic impacts. Hence, normal flood control and FPM measures, while necessary precursors for DRR are not sufficient to deal with low probability-high consequence events that overwhelm the capabilities of local entities to deal with large-scale disasters. Conceptually, federal systems should be able to deal effectively with such events, but a series of recent disastrous floods and storm surges in the USA has exposed some of the weaknesses in the response of federally-based disaster management systems.

(b) Simply put, there are 'many cooks' involved in disaster planning, response and execution in the USA, with mixed lines of authority for execution. Even though the USA has vast resources and organizational capacities to plan for and respond to such large-scale emergencies, the federal system of governance has become complex, with multiple layers of authority and responsibility, making effective coordination before, during and after such events difficult.

(c) As mega-cities grow rapidly in the low-lying coastal areas in the USA, they will become ever-more vulnerable to large-scale events mainly because of the inability to control haphazard land use development. US federal regulations, restrictions and incentives have been inadequate to deal with large urban centers that already exist, such as New York, New Orleans, Los Angeles or Miami. (These problems will be magnified by the sprawling growth of such cities as Dhaka, Manila, Lagos or Jakarta.)

(d) Centralized, coordinated and strategically engineered solutions, such as those undertaken by the Dutch for their coastal protection system, is what is often required for problems of the scale of DRR. The system in the USA that comes closest to such coordinated planning and implementation is the MRT collection of projects built over a century. This includes the Missouri River, Ohio River and Illinois River system of reservoirs, navigation locks and flood control levees. This system of flood control storage and levees protects numerous downstream urban areas, such as St. Louis, Memphis and New Orleans from catastrophic upstream floods.

There are many other insights and observations that can be drawn from an examination of the way in which the US federal system has evolved in dealing with large-scale DRR simply by examining and comparing the responses of three recent disasters. Some of these insights, or 'lessons learned', can be categorized as follows.

6. International lessons

(a) Within the next generation, more than 70% of the global population will be living in vulnerable coastal areas and the large floodplains of numerous rivers. Most of the populace living in these mega-cities will become much more susceptible to comparable consequences that these three events have shown can happen, even in a fairly well-prepared nation with adequate resources.

(b) One of the lessons learned in recent US cases of water-related disasters (and often missed in the literature) is that the preventative solutions to large-scale disasters, of the magnitude of Katrina or Superstorm Sandy, are difficult when approached piecemeal. They require comprehensive solutions of a scale that often can only be implemented at a coordinated federal level.

(c) These three events, for different reasons, taught us that we must better plan for prevention in the growing mega-cities around the globe and not just view them in terms of humanitarian post-disaster responses alone. Planning for disaster prevention is a key strategic decision that requires serious coordinated planning and execution.

(d) These cases, as well as past events, demonstrate that the relationships between managing uncertainties of extreme water events, economic development and social well-being are critical.

(e) The events are an important affirmation of one of the key findings of the United Nations Secretary General's High Level Experts and Leaders Panel (HELP) on Water and Disasters. Expenditures for disaster prevention are actually investments in development and growth.

(f) In each of these events, damages as a percentage of GDP have remained small in places with risk reduction management. This means the events, as large as they were, could be managed in ways that enabled the socio/economic systems to function – something critical to social stability.

(g) Climate change is likely to increase the magnitude and variability of extreme events. Mega-cities will require more infrastructure investment to protect against such events, together with more restrictive land-use controls and zoning to avoid increases in damages.

7. US national lessons

(a) The evolution of the US federal system over the past century has led to greater fragmentation of planning and decision making, resulting in numerous entities, often with different legally delineated objectives, fulfilling their respective responsibilities as part of a comprehensive DRR plan. This fragmentation of authorities and solutions leads to 'brittle solutions' that are only as sound as the weakest link (e.g. Hurricane Katrina).

(b) Historically, US federal water management agencies were able to plan, design and manage large comprehensive systems, which generally performed well, even under extreme catastrophic flood conditions that exceeded the design flood conditions. Contemporary federalism has led to more dispersed decision-making authorities and accountabilities of numerous agencies for flood fighting operations, responses and post-flood recovery. It is difficult for enhanced coordination to overcome the decision and policy discontinuities inherent in the numerous layers of governmental responsibility inherent in a federal system. Consequently, the USA is experimenting with new forms of partnerships.

(c) The US Congress, which represents local constituents, is more likely to favor and fund proven, reliable structural solutions. This often sets up a dichotomy between the federal agencies which advocate and operate under non-structural principles of FPM as part of a sustainable development strategy, and Congress, which wants to ensure the safety of its citizens and is focused more on reliability of solutions rather than 'resilient communities' with relatively high degrees of unquantified levels of 'residual risk'.

(d) Along the Atlantic coast, there was significantly less damage and social disruption from Superstorm Sandy around those areas with existing hurricane shore protection projects. Therefore, USACE has been authorized by Congress to return authorized Hurricane Storm damage reduction projects in nine states along the Atlantic coast to pre-Superstorm Sandy conditions, and repair damages at USACE operating projects and facilities due to Sandy. This work is furthest along of the federal reconstruction efforts.

8. Technical lessons

(a) The best-performing system in recent US cases was the MR&T during the 2011 floods. The system was planned, designed and built over a century, and managed by a single agency – the Corps of Engineers, which was responsible for the project performance during the flood and was accountable for its decisions and failures. This model is unlikely to be replicated in the USA. However, it may be an option in other less developed nations, which are at the same stage of history and development as the USA was in the 1950s.

(b) Large-scale disasters require comprehensive and well-coordinated management plans, with consistent decision rules and performance criteria. New risk-based decision making is complex, and may be placing an unreasonable burden on the public and local officials to sort through and tradeoff various options – each with different degrees of uncertainty and outcomes.

(c) Quantifying risk and uncertainty is at the center of DRR approaches to increasing resiliency, and herein lays a core dilemma. Engineering design standards already incorporate a socially determined 'acceptable' level of social risk tolerance, and they have the virtue of being uniformly applied. Risk-based decision making at the local level is often neither replicable nor uniform. It is not at all clear whether a collection of loosely connected local solutions, with varying degrees of largely unknowable residual risk and uncertainty, can guarantee either robustness or resilience.

(d) Residual risk is almost always underestimated because it is difficult to quantify a cascading series of highly interdependent measures, each of which has its own reliability characteristics and risk of failure. Mostly, though, a flood protection system is a collection of fragmented measures, including new building codes, zoning ordinances and structural measures that are implemented over a long period of time and loosely coordinated by multiple authorities. This comprises the definition of a 'brittle system'.

(e) Storm surge barriers were actually planned but never implemented in New Orleans and NYC in the mid-1970s; today a barrier has been built in Louisiana and talk of such has been renewed in NYC. Conversely, the 2011 Mississippi flood of record showed us how well plans can bear benefits almost 80 years after their implementation.

(f) Because people bear the risks, their involvement in choosing risk and participation in the tough operational decisions made during the process of planning for mitigation of potential events is critical to the health of a democratic system. But this is not easy; leaders are awakened during disasters but quickly forget once disasters subside. Communication of and participation in choosing acceptable risk was mixed with Katrina; it was clear, truthful and transparent in the recent Mississippi floods, but seemed rarely discussed in the case of Sandy.

(g) Rather than being defined purely by engineering solutions, we now understand that all stakeholders contribute to risk reduction through a variety of structural and non-structural measures. This means we must include zoning, building codes, outreach, evacuation plans, warning, insurance and design of structures in risk reduction strategies.

(h) Multiple defenses are critical. This means creating packages of natural and human processes such as evacuation routes, elevated and flood-proofed buildings, pumping stations, levees, flood gates, highways, natural ridges, wetland nourishment, barrier islands, outer shelf activities and compartmentalization of polder areas.
(i) Relief, response and prevention must be integrated in an holistic system of risk management, where citizens actively participate.
(j) In each of these cases, important forensic and decision analysis studies have been initiated. This is critical as they are important in defining the events and responses in technical terms, as much as possible.
(k) Institutional arrangements are necessary for disaster planning, risk reduction and reaction to disasters. Institutional arrangements depend on the political systems and cultures within which they work. The US example is a federal system where many levels of authority coexist.

9. Conclusions

To a large extent, disasters and relief are now globalized phenomena, and we need to rethink how to cooperate in prevention as well as reaction. Urban areas are growing rapidly, often with inadequate flood and storm surge protection. Leaders must plan for prevention of disasters and not just count on humanitarian responses to events. DRR is an important component of economic development, especially in developing nations. Water related DRR programs should be more integrated and combine other uses of water storage, such as supply, with flood defense measures. Integration, however, is becoming exceedingly more difficult as fragmentation in decision making increases exponentially, especially in modern federal systems, with multiple levels of responsibility. Also, ecosystem vitality must be integrated into the solutions. Programs must view the integration of relief, response and prevention as holistic risk management – where citizens actively participate in choosing levels of acceptable risk.

Water related DRR is becoming exceedingly more complex. The USA has considerable technical experience but is overcome by a proliferation of often conflicting policies that are implemented by numerous regulatory agencies with veto power, though with little responsibility for the consequences of their decisions. We have moved from fear, to control, to prevention, to management, and now to working with the floods or disasters. Conventional flood-control strategies have worked well for thousands of communities. DRR requires a higher level of coordinated planning and management to include coordination of structural and non-structural approaches so as to work together. However, most important is changing the behavior of both citizens and decision makers.

Flood management should be an important part of Integrated Water Resources Management (IWRM) and optimal use of water. However, it brings additional complexity. It links land to water and upstream to downstream. Flood management or control is a public good and offers additional revenue sources for IWRM projects. It can also help bring riparians to negotiate around benefits versus allocating flows.

References

American Society of Civil Engineers (ASCE) (2007). *The New Orleans Hurricane Protection System, What Went Wrong and Why, A report by the American Society of Civil Engineers Hurricane Katrina External Review Panel*. ASCE, Reston, Virginia. www.pubs.asce.org.

Arnold, J. L. (1988). *The Evolution of the 1936 Flood Control Act.* Office of History, US Army Corps of Engineers, Fort Belvoir, Virginia.

Association of State Floodplain Managers (ASFPM) (2004). *No Adverse Impact Floodplain Management – Community Case Studies 2004.* ASFPM Foundation, Madison, WI. 72 pp.

Billington, D. P., Jackson, D. C. & Melosi, M. V. (2005). *The History of Large Dams: Planning, Design and Construction.* US Dept of Interior, Bureau of Reclamation, Denver, CO.

Burby, R. J., Bollens, S. J., Kaiser, E. J., Mullan, D. & Sheaffer, J. R. (1988). *Cities Under Water: A Comparative Evaluation of Ten Cities' Efforts to Manage Floodplain Land Use.* Boulder: Monograph #47. Institute of Behavioral Science, University of Colorado, Boulder.

Burton, M. L. & Hicks, M. J. (2005). *Hurricane Katrina: Preliminary Estimates of Commercial and Public Sector Damages.* Center for Business and Economic Research, Marshall University, WV, USA, 12 pp.

Finucane, M. L., Clancy, N., Willis, H. H. & Knopman, D. (2014). *An Initial Assessment of Federal Agencies' Implementation of the Hurricane Sandy Rebuilding Task Force's Infrastructure Resilience Guidelines.* RR-841-DHS Prepared for US Department of Homeland Security, Office of Infrastructure Protection. RAND Corp Homeland Security and Defense Center.

Hessler, W. L. (2000). *The Nation's Response to Flood Disasters: A Historical Account.* Association of State Floodplain Managers, Inc., Madison, WI. www.asfm@floods.org or www.floods.org.

Holliday, W., Floyd, J. & Chao, P. (1998). *Impediments to Evaluation and Development of Non-structural Flood Damage Reduction Measures.* Institute for Water Resources, US Army Corps of Engineers, Casey Building, Ft. Belvoir, Alexandria, Virginia.

Holmes, B. (1972). *A History of Federal Water Resources Program 1800–1960.* US Department of Agriculture, Economic Research Service, Washington, DC, USA.

Kenney, D. (1997). *Resource Management at the Watershed Level: An Assessment of the Changing Federal Role in the Emerging Era of Community-Based Watershed Management.* Report to the Western Water Policy Review Advisory Commission. Natural Resources Law Center, University of Colorado School of Law Boulder, Colorado, 60.

Maass, A. (1951). *Muddy Waters – The US Army Engineers and the Nation's Waters.* Harvard Uni. Press, Cambridge, MA.

Maass, A., Hufschmidt, M., Dorfman, R., Thomas, H. A., Marglin, S. A. & Fair, G. M. (1962). *Design of Water Resources Systems.* Harvard Uni. Press, Cambridge, MA.

Milly, P. C. D., Wetherald, R. T., Dunne, K. A. & Delworth, T. L. (2002). Increasing risk of great floods in a changing climate. *Nature 415*, 514–517. doi: 10.1038/415514a.

Mohleji, S. & Pielke Jr, R. (2008). *Losses from Weather Events: 1980–2008.* doi: 10.1061/(ASCE)NH.1527-6996.0000141. *Natural Hazards Review* 15. American Society of Civil Engineers

Pielke Jr, R. A., Gratz, J., Landsea, C. W., Collins, D., Saunders, M. & Musulin, R. (2008). Normalized hurricane damage in the United States: 1900–2005. *Natural Hazards Review* 10.1061/(ASCE)1527-6988 (2008)9:1(29), 29–42.

Pielke Jr, R. A. & Landsea, C. W. (1998). Normalized hurricane damages in the United States: 1925–1995. *Weather and Forecasting 13*: 621–631.

Rosen, H. & Reuss, M. (eds) (1988). *The Flood Control Challenge: Past, Present and Future.* The Public Works Historical Society, Washington, DC, USA.

Shabman, L. (1988). The Benefits and Costs of Flood Control: Reflections on the Flood Control Act of 1936. In: *The Flood Control Challenge: Past, Present and Future.* Rosen, H. & Reuss, M. (eds). Public Works Historical Society.

Stakhiv, E. (2011). Pragmatic approaches for water management under climate change uncertainty. *JAWRA 47* (6), 1183–1196. doi: 10.1111/j.1752-1688.2011.00589.x.

Stakhiv, E. & Major, D. (1997). Ecosystem Evaluation, Climate Change and Water Resources Planning. In: Frederick, K., Major, D. & Stakhiv, D. (eds), *Climate Change and Water Resources Planning Criteria.* Kluwer Academic Publishers, Dordrecht, pp. 103–118.

Stern, P. & Feinberg, H. (eds) (1996). *Understanding Risk: Informing Decisions in a Democratic Society.* National Research Council, Committee on Risk Characterization, National Academy Press, Washington, DC, USA.

United States Army Corps of Engineers (USACE) (2000). *Planning Guidance.* ER 1105-2-100. US Government Printing Office, Washington, DC, USA, 22 April 2000.

United States Army Corps of Engineers (USACE) (2014). *North Atlantic coast comprehensive study: resilient adaptation to increasing risk.* Main Report; Final Report November, 2014. USACE North Atlantic Division.

United States Army Corps of Engineers, Institute for Water Resources (USACE, IWR) (2008). *Decision Making Chronology for the Lake Pontchartrain & Vicinity Hurricane Protection Project*. Institute for Water Resources, US Army Corps of Engineers, Ft Belvoir, Virginia. http://library.water-esources.us/docs/hpdc/hpdc.cfm (Revised 10 October 2008).

United States Army Corps of Engineers (USACE), Mississippi Valley Division (MVD) (2008). Designing the Project Flood. http://www.mvd.usace.army.mil/Portals/52/docs/Designing%20the%20Project%20Flood%20info%20paper.pdf.

United States Army Corps of Engineers (USACE), Mississippi Valley Division (MVD) (2012). *Post 2011 flood report*. http://www.mvd.usace.army.mil/Missions/FloodRiskManagement/RegionalFloodRiskManagementProgram/MRTPostFloodReport.aspx.

United States Army Corps of Engineers (USACE), Mississippi Valley Division (MVD) (2014). Mississippi River and Tributary Project. http://www.mvd.usace.army.mil/About/MississippiRiverCommission(MRC)/MississippiRiverTributariesProject(MRT).aspx.

United States Congressional Research Service (CRS), Congressional Budget Office (CBO) (2008). *Infrastructure Spending Supplemental Tables*. CBO, Washington, DC, USA. http://www.cbo.gov/doc.cfm?index=9135.

United States Congressional Research Service (CRS) (2013). *Analysis of the Sandy recovery improvement act of 2013*, CRS Report for Congress, 11 March 2013. CRS, Washington, DC, USA.

United States Congressional Research Service (CRS) (2014). *Achieving Disaster Resilience in US. Communities: Executive Branch, Congressional and Private Sector Efforts*. Stephanie Sanok Kostro and Garrett Riba. CRS, Washington, DC, USA.

United States Council on Environmental Quality (CEQ) (2014). *Updated Principles and Guidelines for Water and Land Related Resources Implementation Studies*. CEQ, Washington, DC, USA. http://www.whitehouse.gov/administration/eop/ceq/initiatives/PandG.

United States Federal Emergency Management Agency (FEMA) (2000). *National Flood Insurance Program*. FEMA, Washington, DC, USA. www.fema.gov/nfip/qanda.htm.

United States Federal Emergency Management Agency (FEMA) (2014). *A Unified National Program for Floodplain Management*. FEMA, Washington, DC, USA. https://www.fema.gov/media-library/assets/documents/18472.

United States Federal Interagency Floodplain Management Task Force (1992). *Floodplain Management in the United States: An Assessment Report, FIA-18*. Federal Emergency Management Agency, Washington, DC, USA.

United States Federal Interagency Management Task Force (1994). *A Unified National Program for Floodplain Management 1994*. Federal Emergency Management Agency, Washington, DC, USA.

United States Federal Task Force on Flood Control Policy (1966). A Unified National Program for Managing Flood Losses. In: *House Document No. 465: 89th Congress, 2nd session, committee on Public Works*, US Government Printing Office, Washington, DC, USA.

United States Government, General Accounting Office (GAO) (2014). *Emergency transportation relief: agencies could improve collaboration begun during Hurricane Sandy response*. Report to the Chairman, Subcommittee on Transportation, Housing and Urban Development, and Related Agencies, Committee on Appropriations, House of Representatives, Washington, DC, USA. GAO-14-512.

United States Interagency Floodplain Management Review Committee (1994). Report to the President. In: *Sharing the Challenge: Floodplain Management into the 21st Century*, Washington, DC, USA. June 1994.

United States Interagency Performance Evaluation Task Force (IPET) (2009). *Performance of the New Orleans and Southeast Louisiana Hurricane Protection System final Report of the Interagency Performance Evaluation Task Forces*. USACE, Washington DC, USA. http://biotech.law.lsu.edu/katrina/ipet/Volume%20I%20FINAL%2023Jun09%20mh.pdf.

United States National Climate Assessment (2014). *Climate Change Impacts in the United States*. National Climate Assessment, US Global Change Research Program, Washington, DC, USA.

United States, NOAA & USACE (2013). Infrastructure Systems Rebuilding Principles. 28 February 2013. http://coast.noaa.gov/digitalcoast/sites/default/files/files/publications/04062013/InfrastructureSystemsRebuildingPrinciples.pdfLC.

United States Office of Management and Budget (1998). *Federal Programs Offering Non-Structural Flood Recovery and Floodplain Management Alternatives*. Executive Office of the President, Washington, DC, USA.

United States Water Resources Council (1973). Principles and Standards for Planning Water and Related Land Resources. *Federal Register* 38 (174): 24788. Washington, DC, USA.

United States Water Resources Council (1976). *A Unified National Program for Floodplain Management*. Government Printing Office, Washington, DC, USA.

United States Water Resources Council (1983). *Economic and Environmental Principles and Guidelines for Water and Related Land Resources for Implementation Studies*. Government Printing Office, Washington, DC, USA.

Weiner, J. (1996). *The Socioeconomic Aspects of Flooding in the US*. Natural Hazards Research and Application Information Center, University of Colorado, Boulder, www.colorado.edu/hazards/tb19.html.

White, G. (1945). *Human Adjustment to Floods: A Geographical Approach to the Flood Problem in the US*. Research Paper No. 29, University of Chicago, Chicago, Illinois.

White, G. (1969). *Strategies of American Water Management*. U. Michigan Press, Ann Arbor.

A tale of three cities: water disaster policy responses in Bangkok, Kuala Lumpur and Metro Manila

Keizrul Abdullah[a], Apichart Anukularmphai[b], Tadashige Kawasaki[c] and Dolora Nepomuceno[d]

[a]*Corresponding author. Network of Asian River Basin Organizations (NARBO), Department of Irrigation and Drainage, Kuala Lumpur, Malaysia. E-mail: keizrul@gmail.com*

[b]*NARBO and Water and Flood Management Commission, Bangkok, Thailand*

[c]*NARBO, Japan Water Agency, Saitama City, Japan*

[d]*NARBO, Laguna Lake Development Authority, Manila, Philippines*

Abstract

Floods and typhoons are two of the greatest water disasters affecting South East Asia, causing misery and death to people, damaging properties, infrastructure and crops, and causing disruption to commerce and industry. In many cases the impact can be widespread, affecting not only individual households but also large parts of a country including agriculture areas, towns and cities, and sometimes even beyond national borders. The rapid pace of development has resulted in a disproportionate increase in runoff and a many-fold increase in river discharges leading to more frequent and more intense flooding. This situation is expected to be further aggravated due to the impact of global warming and climate change. To cope with such challenges, countries in South East Asia are developing their policy responses tailored to suit their local conditions and environment. This paper looks at the water disaster situation and the policy responses in three cities in South East Asia: Bangkok, Kuala Lumpur and Metro Manila, the capital cities of the Kingdom of Thailand, the Federation of Malaysia and the Republic of the Philippines, respectively. Although all three countries are in the same climatic zone, due to their geographical locations, water disasters impact differently on them and the remedial measures also differ.

Keywords: Policy response; South East Asia; Water disaster

1. Introduction

Floods and typhoons are two of the greatest natural disasters affecting mankind, causing misery and death to people, damaging properties, infrastructure and crops, and causing disruption to commerce and industry. In many cases the impacts can be widespread, affecting not only individual households but also large parts of a country including agriculture areas, towns and cities, and sometimes even

doi: 10.2166/wp.2015.005

beyond national borders. As an example, the Great Flood of Bangkok in 2011 affected the global supply chains for automobiles and computer hard disks. Floods are also a threat to a country or region achieving its full economic development potential, as investors are deterred from venturing into flood-prone areas.

The rapid pace of development in the ASEAN[1] region has resulted in a disproportionate increase in runoff (that portion of rain that flows overland to a river) and a halving of the time of concentration (the time it takes for the runoff to reach the river). The consequence of this is a many-fold increase in river discharge leading to more frequent and more intense flooding (Figure 1). This situation is expected to be further aggravated due to the impact of global warming and climate change. The Intergovernmental Panel on Climate Change (IPCC) has assessed that there will be possible intensification of the hydrological cycle on a global basis, leading to increased frequencies and intensities of extreme weather events, i.e. floods and droughts will occur more frequently and will be larger and more intense and last longer. In the words of the UN Secretary General, Ban Ki-Moon, 'The abnormal is the new normal.'

The traditional engineering approach to managing floods has been to try to 'control' the flood through structural measures. These have included the construction of flood storage dams and flood attenuation ponds, the deepening and widening of rivers to increase their capacities, and the protection of low-lying areas through the provision of river levees. Such an approach tends to be problem driven, where projects are implemented to reduce the risk of flooding in an area without giving much consideration to the impact of the projects on upstream or downstream areas. In many cases, such piecemeal solutions end up transferring the flood problem to another area.

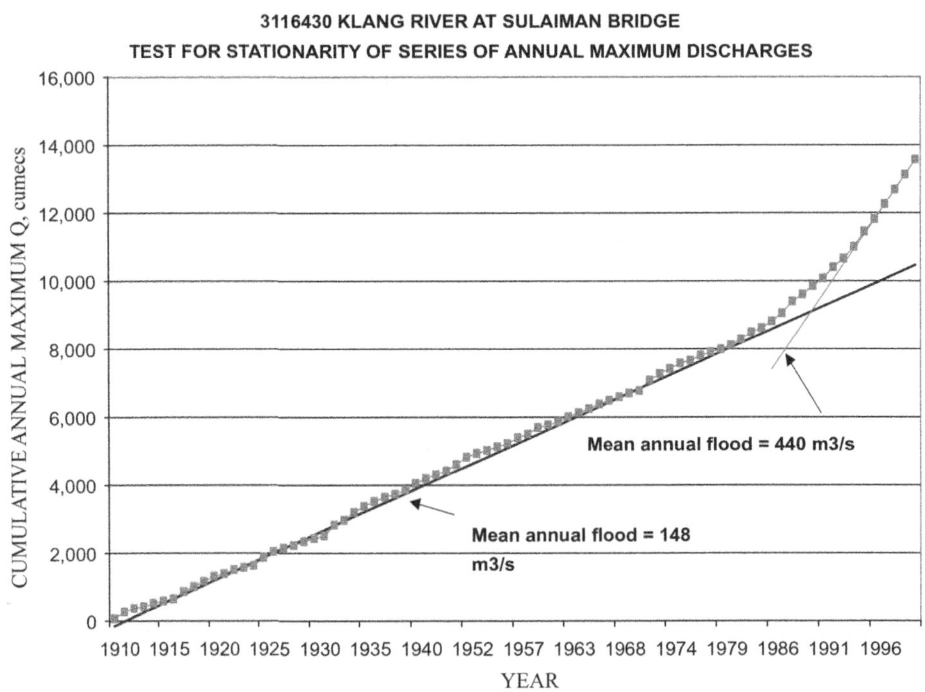

Fig. 1. A three-fold increase in the mean annual flood of Klang River as a result of development of the upper catchment (Abdullah, 2006).

[1] Association of South East Asia Nations.

In recent years, flood engineers have modified the approach to 'mitigate' (reduce the impact) rather than attempting to 'control' floods. This approach involves a combination of structural and non-structural measures. Non-structural or 'soft' measures include flood zoning, flood forecasting and warning, flood risk mapping and flood insurance. However, such measures have proven to be inadequate as flood mitigation works have been unable to keep pace with the rapidly increasing flood volumes arising from development and urbanization.

The challenge now is to focus on 'adaptation' (adjusting to changing situations), i.e. flood solutions have to be innovative, flexible and adaptable to changing situations. At the same time the solutions have to be viewed at a macro level and in the context of a river basin, so as to promote a holistic and integrated approach rather than being localized and fragmented. Such an approach integrates land and water resources development in a river basin and aims at maximizing the benefits to society while minimizing the damage from flooding.

2. Three cities, one region

The Association of South East Asian Nations (ASEAN) is a political and economic grouping of 10 countries located in Southeast Asia, and was formed on 8 August 1967 by Indonesia, Malaysia, the Philippines, Singapore and Thailand. Since then, membership has expanded to include Brunei, Cambodia, Laos, Myanmar and Vietnam (Figure 2). Its objectives include accelerating economic growth, social progress, socio-cultural evolution among its members, protection of regional peace and stability, and opportunities for member countries to discuss differences peacefully.

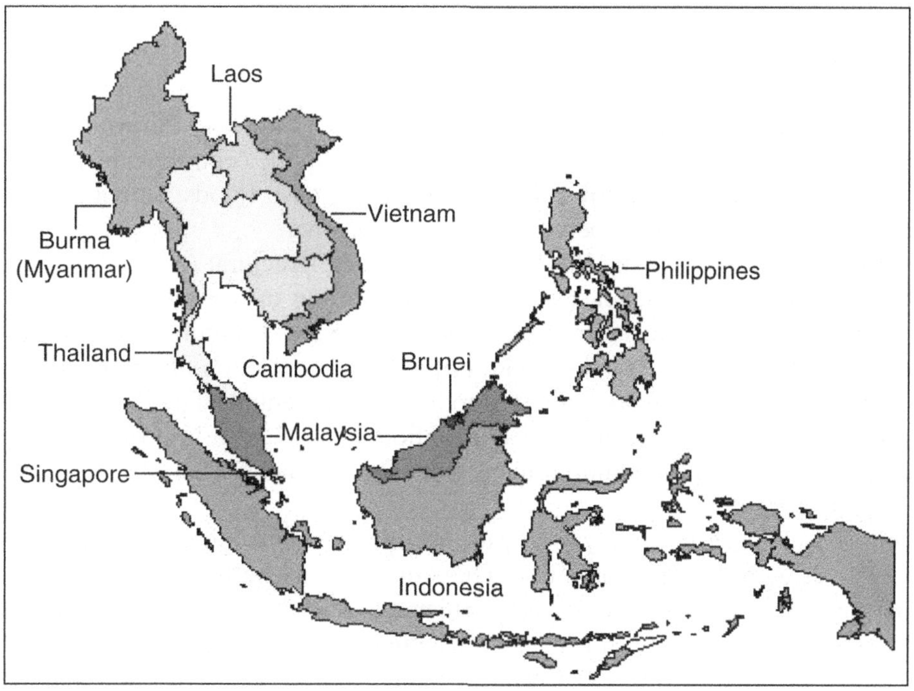

Fig. 2. The 10 countries of ASEAN (see http://www.asean.org/).

ASEAN covers a land area of 4.46 million km^2, which is 3% of the total land area of the Earth, and has a population of approximately 600 million people, which is 8.8% of the world's population. In 2012, its combined nominal gross domestic product (GDP) had grown to more than US$2.3 trillion, which if ASEAN was a single entity, would rank it as the sixth largest economy in the world.

Bangkok, Kuala Lumpur and Metro Manila are the capital cities of Thailand, Malaysia and the Philippines, respectively. The city of Bangkok is located in the estuary of the Chao Phraya River which flows 372 km through the central plains of Thailand into the Gulf of Thailand, draining an area of 160,400 km^2. The city of Kuala Lumpur was founded in the late 19th century as a tin mining settlement and is situated in the middle reach of the Klang River which originates from the central mountain range at an elevation of 1,330 m and traverses a distance of nearly 120 km before discharging into the Straits of Malacca, draining a catchment of about 1,288km^2. Both cities suffer from flooding due to overflow of the Chao Phraya and the Klang rivers, respectively. Metro Manila which is located on the island of Luzon, sits astride the typhoon belt and experiences heavy torrential rains during the typhoon season from July to October.

3. Policy response to water disasters

Climate-wise, ASEAN lies in the humid tropics, defined as a zone located between latitude 23.5° north and 23.5° south and having monthly precipitation exceeding 100 mm per month for at least 4½ months. In fact, most of ASEAN lies in the monsoonal humid tropics, where more than half the year is wet and the annual rainfall ranges from 1,500 to 5,000 mm. As a result, floods are common during the monsoonal wet months.

3.1. Policy response – Bangkok

Bangkok is located in the estuary of the Chao Phraya River. The Chao Phraya begins at the confluence of the Ping and Nan rivers and flows south for 372 km from the central plains to Bangkok and into the Gulf of Thailand, draining an area of 160,400 km^2. In the past, floods were a common occurrence especially during the monsoon season between May and October.

The policy response to deal with the floods was very much structural (or engineering) based, i.e. to store the flood waters upstream and release them for use during the dry season. As part of the flood management plan, three large dams were constructed: the King Bhumibol Dam (1964), the Queen Sirikit Dam (1972) and the Pasak Chonlasit Dam (1998). On completion, the three dams provided sufficient storage capacity to substantially reduce flooding in the central plain and the city of Bangkok. With the floods considered under 'control', less attention was given to improving the capacity of the Chao Phraya River downstream. In addition, many drainage channels in the city were deemed superfluous and converted into roads.

The monsoon of 2011 was unusually wet with accumulated precipitation exceeding the historical average by more than two-fold in some areas. The situation was further exacerbated by four tropical storms and one typhoon that hit the north and northeast late in the rainy season when normally there would be only two to three tropical storms. This resulted in the total amount of rainfall in the upper Chao Phraya River Basin being 30–60% higher than average. Consequently, the water levels in the three reservoirs not only reached historical highs but peaked some 2 months early, resulting in the need to release the excess flood waters downstream. This huge volume of water created a flood wave downstream which lasted for over 2 months.

The weakness in the flood management plan was that it could not cope with the very extreme series of hydrological events that occurred, and when the storage capacities of the three dams were exceeded, there was no back-up plan downstream. In the light of the IPCC's assessment that there will be possible intensification of the hydrological cycle leading to increased frequencies and intensities of extreme weather events, the 'abnormal' flood of 2011 may become the new 'normal' in the future.

Subsequent to the 2011 floods, the policy response of the Thai government has been to focus on three areas: infrastructure, management and climate change. In the area of water infrastructure, existing design concepts and criteria are being reviewed and upgraded, while the government has begun a huge programme estimated to cost around 100 billion[2] bahts to repair, rehabilitate and upgrade the flood infrastructure. In terms of management, nature can no longer be taken for granted and 'business as usual' is no longer tenable. The planning process will now consider the entire spectrum of upstream to downstream factors and include consultations with all stakeholders to produce an over-arching flood and water management plan that mitigates, as far as possible, a recurrence of the 2011 catastrophic flooding.

To cope with changing climatic conditions, Thailand is looking to develop technologies in predicting/forecasting weather more precisely so that there will be sufficient time to take preventive and protective measures, and provide early warning to the population. For the long-term, people will have to learn to live under extreme hydrological events and adjust by building resilience capabilities in their communities and infrastructure.

3.2. Policy response – Kuala Lumpur

Kuala Lumpur is situated at the confluence of two tributaries of the Klang River and has, from the earliest days, been subjected to flooding, with the 1926 flood being the worst ever recorded. The Klang River is small by global standards, having a length of 120 km and a catchment area of only 1,288 km^2. The river's headwaters comprise mountainous and steep terrain and, coupled with the intense level of precipitation, this results in large volumes of storm runoff even from small sub-catchments, which often leads to the river systems in the localized areas being completely overloaded. This causes what is termed as 'flash' flooding which is quick to manifest and equally swift to subside. This form of flooding is frequently experienced and can occur several times in a year. In addition, as a result of the extensive urbanization of the catchment up to the hill slopes, the severity and frequency of floods have increased over the years.

In the early days, the policy response was to 'control' floods through structural (engineering) measures, and a programme to channelize and dredge the river system was undertaken. However, from the 1970s, as Kuala Lumpur continued to transform into the ultra-modern metropolis it is today and as development kept changing the face of the city, the existing infrastructures came regularly under pressure to cope with flood flows beyond their design limits. This urbanization of the catchments in and around the city has resulted in flood runoffs increasing many fold, bringing with it challenges for the flood planners to come up with more innovative solutions.

---

[2] Billion = 10^9.

One such solution was to construct a flood bypass to divert the flood flows away from the city centre. However, the topography of the land was such that a bypass alignment would cut through high ground as well as areas which were densely built-up, making the use of an open channel (open cut) prohibitive in terms of land acquisition and compensation costs. It was thus decided that the bypass should be a tunnel as this was more competitive in terms of pricing and would offer the least disturbance during construction. At the same time, it was realized that a flood bypass tunnel would be in use infrequently and hence by incorporating a road into the flood tunnel, value could be added to the project by introducing an additional function of helping to reduce the traffic congestion at the southern gateway of the city into the city centre.

The resulting Stormwater Management and Road Tunnel (SMART) project is the first project in Malaysia (and possibly in the world) to combine two contradicting uses in one tunnel, i.e. to pass flood waters and traffic through the same tunnel, albeit at different points in time. It was also funded through a new financing mechanism whereby the costs were shared between the government and the private sector, with public funds paying for the flood component and the private sector funding the motorway component. The private sector was given a concession to recover costs by charging a toll on motorway users and, in the process, help to defray public costs in operation and maintenance.

3.3. Policy response – Metro Manila

Sitting astride the typhoon belt and having a long coastline of 36,289 km, the Philippines archipelago is particularly vulnerable to tropical storms and typhoons. On average, up to 20 tropical storms enter Filipino waters annually, with about half of them making landfall. The most frequently impacted areas are Luzon Island and the eastern Visayas. The capital, Metro Manila, is situated on the island of Luzon and is ranked by the USA as the third most vulnerable metropolitan area on earth with the second most number of people at risk.

To some extent, Metro Manila is fortunate because of its geographical positioning whereby the rugged terrain surrounding the city has, in the past, effectively served as a windbreak. Two previous typhoons with wind speeds of more than 110 knots (200 km/h) fell to 75 knots (140 km/h) by the time they reached the city.

Typhoons bring heavy rainfall which, when combined with the strong winds and storm surges, can cause large-scale flooding and result in enormous destruction and loss of life. In the face of such daunting factors, it would be prohibitively costly to invest in the necessary defensive infrastructure to protect Metro Manila from typhoons. Instead, the policy response has been to focus on improving disaster preparedness and management.

Unfortunately, so far, city officials have shown a disheartening inability to cope with such disasters. In 2012, torrential rains hovered over the city for 8 days, causing citywide flooding, leaving 63 people dead and nearly 84,500 homes destroyed completely. Despite the fact that the storm hit the centre of government, it was weeks before the city could recover from the damage and destruction.

In the aftermath of Super Typhoon Haiyan, funding for disaster preparedness and management was increased by nearly 100 billion Pesos, or about 5% of the total national budget. However, while a portion has been spent wisely on more proactive and innovative disaster mitigation programmes such as hazard mapping, the bulk of the funds is earmarked for cleaning up, repairs and rehabilitation of the damage from Haiyan.

4. Case studies

4.1. The 2011 flood: Thailand's worst flood

4.1.1. Introduction. The Kingdom of Thailand is a country at the centre of the Indochina peninsula. It is bordered to the west by Myanmar, to the east by Cambodia and Laos, and to the south by Malaysia. Located in the humid tropics, Thailand has a total land area of approximately 513,000 km^2 and a population of around 65 million people. The capital and largest city is Bangkok, which is Thailand's political, commercial, industrial and cultural hub. Over the past three decades, Thailand has experienced rapid economic growth and is now considered as a newly industrialized country.

The average annual rainfall in Thailand ranges from a high of almost 4,000 mm in the south to a low of 1,000 mm in the northern and central plains. The monsoon season takes place between May and October and the heavy rainfall during this period is a result of low pressure cells, depression storms, tropical storms and typhoons originating in the South China Sea and moving towards the Indochina region.

In the Chao Phraya River Basin, the average annual rainfall ranges from 966 to 1,259 mm per year (based on the past 30 years' record). In 2011, the accumulated rainfall from January to November was 1,781 mm, which was approximately 30% higher than the average value. Table 1 gives the annual rainfall accumulation in Chao Phraya River Basin since 1978.

4.1.2. The 2011 floods. The 2011 floods were the worst to ever hit Thailand. The flooding of the central plains of Thailand started in the month of June and progressed downwards towards Bangkok in the following sequence of events. During the months of March and May, Thailand had experienced a heavy monsoon season which was characterized by high rainfall. At the end of June, tropical storm Haima brought heavy and widespread rainfall to the northern part of Thailand causing flash flooding in five provinces including Chiang Rai, Phayao, Nan, Tak and Sukhothai.

In early August, tropical storm Nock-Ten hit the northern region bringing heavy rainfall to Nan, Uttaradit, Phrae and Pitsanulok provinces. By mid-August, the floods had reached Nakhon Sawan Province while the flooding situation remained critical in the provinces of Chiang Rai, Lamphun, Lampang, Sukhothai, Pitchit, Phitsanulok, Kamphaeng Phet and Nakhon Sawan.

By early September, the floods had reached Sing Buri, Chainat and Lopburi. Two weeks later the floods reached Ayutthaya. By early October, all of Ayutthaya province including Wat Chai Wattahanaram (a World Heritage site) was declared as a disaster area. The flood wave next reached Nonthaburi, causing breaches in the Hitech Industrial Estate dyke. A week later, the flood waters reached the Bangpa-In Industrial Estate and the Nava Nakhon Industrial Estate. On 25 October, the flood waters reached Don Muang Airport and the Flood Relief Operation Centre (FROC) which had to be evacuated and relocated to the Energy Centre. By early November, the flood waters had advanced

Table 1. Average annual rainfall in Chao Phraya River Basin since 1978 (in descending order of magnitude) (*Source:* Anukularmphai, 2014).

Year	2011	1980	1995	1978	1983
Average annual rainfall (mm)	1,781	1,145	1,161	1,151	1,129

to Lad Prao intersection, Lad Prao Road, Rama II Road and Ratchadapisek Road (near Bangkok's Ratchadapisek Metropolitan Rapid Transit (MRT) station).

Finally, on 17 November, the flood waters on Lad Prao Road, Ratchadapisek Road and near-by areas began to recede and by 28 November, the Inner Bangkok area was back to normal, with businesses and shopping malls reopened, except in the Lak Si, Don Muang, Phasi Chareon, Tawiwattana and Bangplad areas. Figure 3 shows the flood wave at various points along the Chao Phraya River.

From Radarsat images captured in the Chao Phraya Basin, it can be seen that at the early stages of flooding in July, the flooded areas were localized and scattered. Then, as flooding progressed towards the end of September, most of the flood waters were concentrated in the upper and lower Chao Phraya. Finally, it moved downwards as a huge mass of flood water towards Bangkok. It has been estimated that about 15,260 million m^3 of flood waters were accumulated above Bangkok in October. This huge volume of water had to be drained into the Gulf of Thailand, mainly through the Chao Phraya River

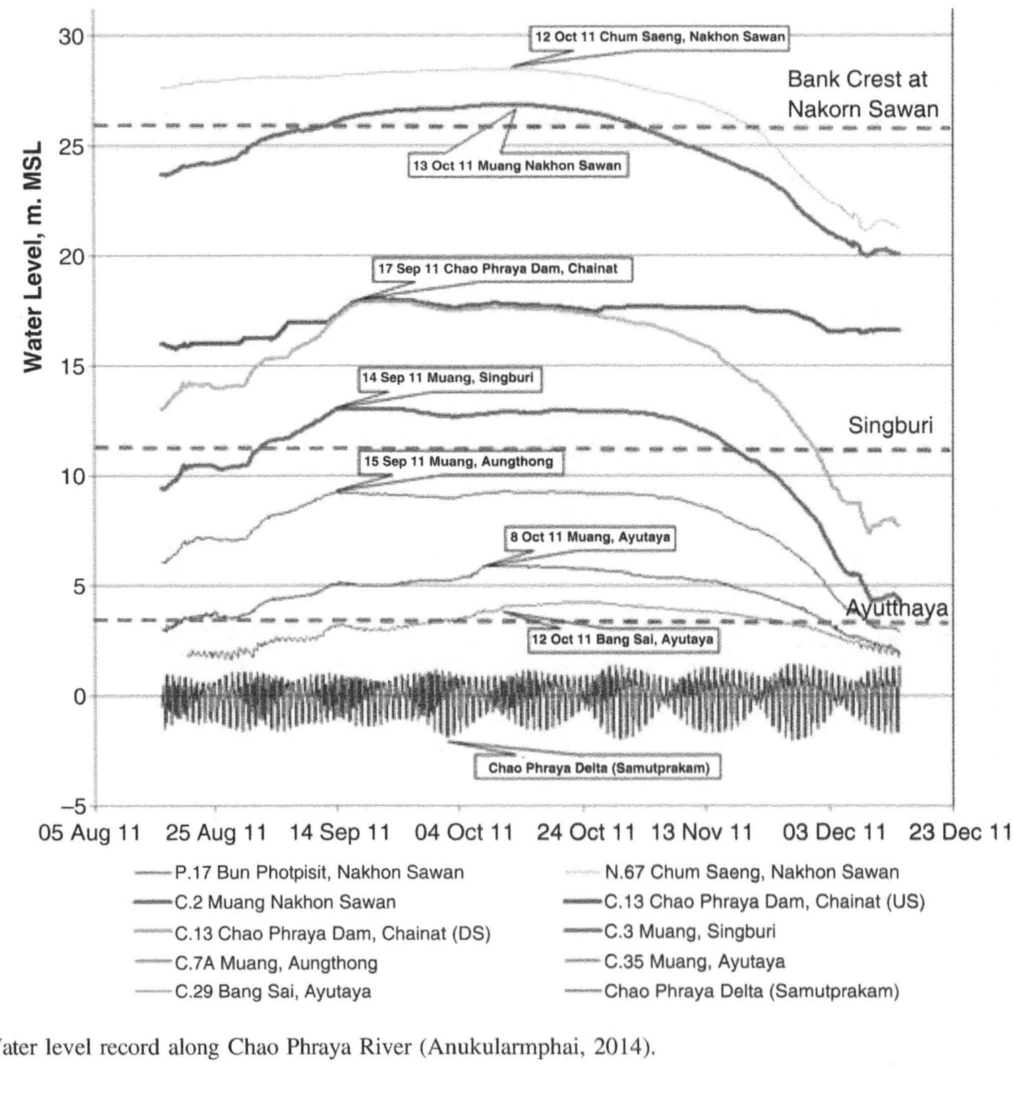

Fig. 3. Water level record along Chao Phraya River (Anukularmphai, 2014).

and its flood plain. The maximum combined outflow of the Chao Phraya river (300 million m^3/day at the estuary) and the estimated drainage system capacity located on both sides of Chao Phraya (200 million m^3/day) was only 500 million m^3/day, which was relatively small in comparison to the total volume of flooded water. This was the basic reason why the flooding situation was prolonged for months in some of the most severely flooded areas.

The main cause of the floods was an unusually wet monsoon exacerbated by four tropical storms (Haima in June, Nock-Ten in August, Haitang in September and Nalkae in October) and one typhoon (Nasat in September) that reached the north and northeast of Thailand late in the rainy season. These storm events were unusual as normally there would be only two to three tropical storms per year. Preceding the floods, Thailand had already experienced a monsoon season which was characterized by high rainfall during the months of March and May, with accumulated precipitation exceeding the historical average value by 277% and 45%, respectively. Coupled with the precipitation from the four storms and typhoon Nasat, the total amount of rainfall in the upper Chao Phraya River Basin was 30–60% higher than the average (Figure 4). In the Bangkok area, the accumulated precipitation up to 1 December 2011 was 2,257.5 mm, while in comparison, in the past 50 years, the highest precipitation level recorded was only 1,973.5 mm.

The floods of 2011 affected 65 of the 76 provinces, 684 districts, 4,920 tambon and 43,636 villages and left heavy extensive damage along their path. In terms of population, 13,595,192 people from 4,086,138 households were affected, with 693 persons confirmed dead and 3 missing. The causes of death were drowning (580 persons), electric shock (48 persons), washed away by the floods (26 persons), boat sinking (23 persons) and others (16 persons). In an effort to cope with the flood-affected population, 1,739 evacuation shelters and stations were opened.

The negative impact on agriculture was significant with some 12.61 million hectares of farm land flooded. For the industrial sectors, 9,859 manufacturing plants with 600,000 factory workers were affected, with the flood damage estimated at 800,000 million baht. At least seven industrial estates

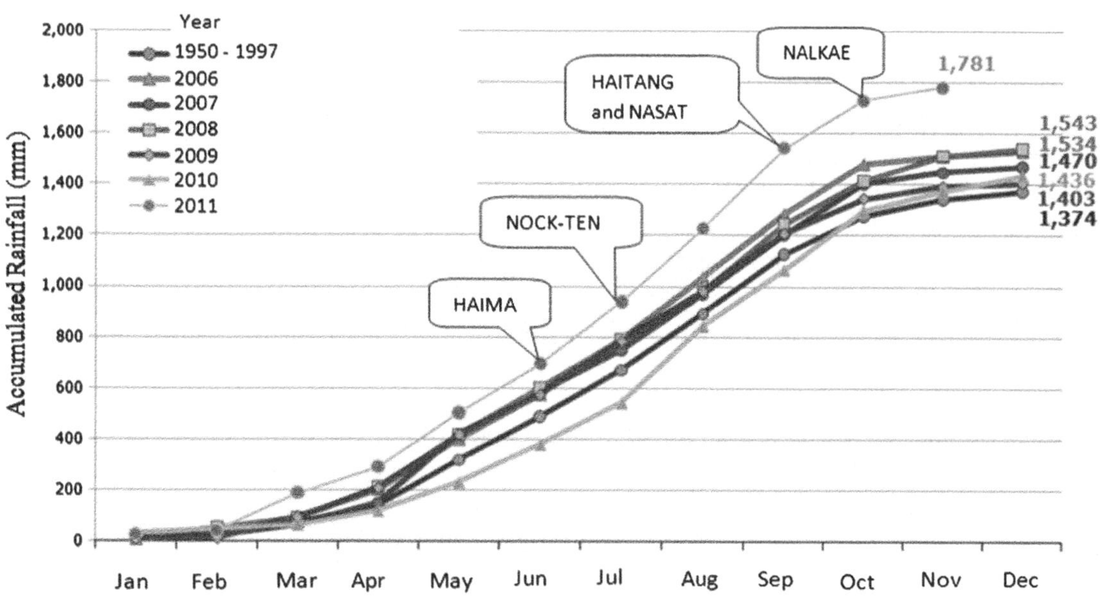

Fig. 4. Accumulated rainfall and major storm events in Thailand in 2011 (Anukularmphai, 2014).

were flooded, damaging 838 factories, affecting 382,693 factory workers and resulting in an estimated flood related cost of 403,784 million baht. The floods' impact on academic institutions such as schools and universities was unprecedented with 3,088 schools disrupted and approximately 700,000 students affected.

In terms of culture and tourism, 313 historical ancient monuments and sites were damaged. In Ayutthaya province, home to numerous world heritage sites, 130 locations were affected. The economic losses for the tourism sector were estimated at 50,000 million baht, due to a 10% decline in tourist arrivals during and after the floods. The World Bank estimated the total economic damage at 1.4 trillion baht (approximately US$44.5 billion), slashing economic growth by 2.4% and making the 2011 Thailand floods the world's fourth costliest disaster, surpassed only by the 2011 earthquake and tsunami in Japan, the 1995 Kobe earthquake and Hurricane Katrina in 2005.

In addition, the impact of the floods was felt beyond the national borders, causing disruption to the global supply chain. Two particularly hard hit sectors were the automobile and computer manufacturing industries. In the automobile industry, two of the world's largest car manufacturers, Honda and Toyota, had severe disruption to their manufacturing processes, not just in Thailand but worldwide, due to a lack of parts. In the computer industry, Thailand was responsible for the manufacturing of about a quarter of all hard disk drives produced for the global market and the flooding of the factories resulted in a global supply shortage.

4.1.3. Policy response: pre-2011 floods. Being in the humid tropics, the need to deal with floods especially during the wet monsoon season, has long been recognized in Thailand. The policy response beginning in the 1950s was very much engineering biased, i.e. to construct three large dams: the King Bhumibol, the Queen Sirikit and the Pasak Chonlasit, to store the flood waters and release them for use during the dry season.

The Bhumibol Dam is a concrete arch-gravity dam on the Ping River, a main tributary of the Chao Phraya River. The dam is 154 m high, 486 m long and 8 m wide at its crest, and has an active storage capacity of 9,762 million m^3. Construction began in 1958 and was finished in 1964 while the reservoir was completely filled in 1970. Located about 480 km north of Bangkok, the dam was built for the purposes of water supply and irrigation, hydroelectric power production, flood control, fisheries and saltwater intrusion management, and was Thailand's first multi-purpose dam project.

The Sirikit Dam, completed in 1972, is an embankment dam on the Nan River, the other main tributary of the Chao Phraya River. Built for the purpose of irrigation, flood control and hydroelectric power production, the dam is 113.6 m high, 800 m long 12 m wide at the crest and 630 m wide at the base. Covering a surface area of 259 km^2, the dam reservoir has an active storage capacity of 6,666 million m^3. Together, the Bhumibol and Sirikit Dams control 22% of the Chao Phraya's annual runoff and both dams also help provide for the irrigation of 1,200,000 ha of farmland in the wet season and 480,000 ha in the dry season.

The Pasak Chonlasit Dam is a 36.5 m high and 4,860 m long earthfill embankment dam with an impervious core, impounding the Parsak River. With a storage capacity of 785 million m^3 at normal water level, and a maximum capacity of 960 million m^3, it is the biggest reservoir in Central Thailand. Commissioned in 1998, the Pasak Chonlasit Dam project is one of Thailand's major irrigation projects, providing water to the plantations in the Pasak valley and the lower Chao Phraya valley. The dam also helps to decrease the problems of water management in Bangkok by allowing more flood control, as the Pasak River is one of the main sources of flooding in the Bangkok metropolitan area.

Collectively, the three dams act together to prevent or reduce the flooding of the lower Chao Phraya flood plain, including the city of Bangkok. Hence, the procedures and operations of the dams during the raining season are of paramount importance for reducing the flood risk to Bangkok. Over the years, a set of operation rule curves have been developed and prior to 2011, these rule curves managed to control the peak inflows within the capacities of the dams. However, the heavy rainfall and the unusual rainfall pattern of 2011 resulted in the dams reaching their maximum capacities some 2 months early, resulting in the need to release the excess flood waters downstream. Figure 5 shows the situation for Bhumibol Dam.

4.1.4. Policy response: during the 2011 floods. As the magnitude of the floods worsened, the government set up a Centre for Emergency Management and mobilized flood monitoring and relief operations. On 7 October 2011, a FROC was established to coordinate the delivery of aid. The FROC included all Ministries, the Bangkok Metropolitan Administration, the Army and other government agencies. It was led by the Minister of Justice who reported directly to the Prime Minister. The FROC administration consisted of two units, an operation team under the Minister of Science and Technology, and a planning and prevention team under the Minister of Transportation. After the FROC had successfully operated within Bangkok, similar organizations were established at the local level, led by the Provincial Governor.

On 20 October 2011, the Prime Minister ordered measures for a designated disaster area, as empowered under the Public Disaster Prevention and Mitigation Act (2007), for 21 provinces that

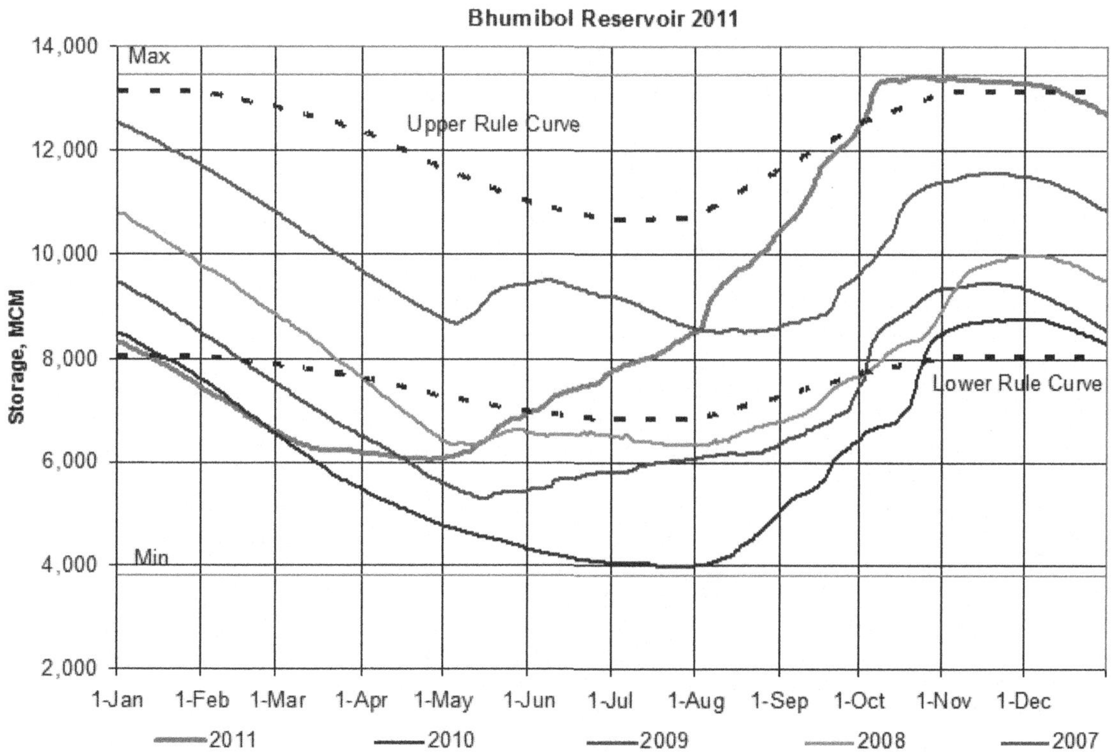

Fig. 5. Operation rule curve of Bhumibol Dam (Anukularmphai, 2014).

were designated as disaster areas. On 25 November 2011, the government declared natural disaster mitigation and relief areas representing 58 provinces and Bangkok. In all, eight committees were established and charged with the various responsibilities for disaster relief and mitigation, as indicated in Table 2.

In general, the government response during the floods could be characterized as somewhat adequate though some of the actions and decisions were reactive rather than proactive. The authority and roles of the numerous committees and working groups varied greatly due to timing, political forces and public responses. A significant gap was the lack of system integration and unity of command, as the various committees and working groups followed their traditional roles and responsibilities according to the legal framework under normal circumstances or non-disaster operating environments.

External and internal public communication were observed to be seriously challenged, especially when flood information requests became more complicated and increasingly area specific. Conflicting, and at times misleading, flood information was a normal occurrence with various agencies offering technical information that was either inaccurate or simply not given in a timely fashion. Consequently, there were several conflicts and incidents of protest in different forms including violence as the public took matters into their own hands to handle the floods in the best possible way that they thought was correct.

Overall, a lack of awareness and preparedness for natural disasters such as floods appeared to be one of the core challenges cutting across government, the private sector and civil society as a whole. As a result of this general lack of public awareness, housing and businesses had encroached into the flood plains at an unprecedented rate with little preparation or preventive thought for flooding or the effects thereof.

Table 2. Establishment of committees and their responsibilities (*Source:* Anukularmphai, 2014).

Name of committee	Responsibility	Date of establishment
1. National committee on floods, tropical storm and mudslides	Oversee and manage national disasters due to floods, tropical storms and mudslides Preparedness, prevention, response and recovery of all the major phases of disaster management	25 August 2011
2. Twenty-five river basin national joint-committee	Support national water management and flood warning system	11 September 2011
3. Flood Relief Operation Centre (FROC)	Oversee all flood management effort focussing on the lower Chao Phraya plain	7 October 2011
4. Flood relief committee (structural, economic, social subcommittee)	Assist the population affected by flood disaster	12 October 2011
5. Flood recovery and restoration committee	Replacement of the flood recovery and restoration committee To coordinate the flood relief management for structural, economic and industrial issues, and people's livelihoods and quality of life	4 November 2011
6. Public communication committee	To improve public information of flood disaster to people	4 November 2011
7. Strategic committee for reconstruction and future development	Prepare strategies for reconstruction and national future of sustainable development	10 November 2011
8. Strategic committee for water resources management	Manage water resources with principles of sustainability	10 November 2011

4.1.5. Policy response: post 2011 floods. The flooding of the central plains of Thailand, including Bangkok, began in late July 2011 and finally reached Bangkok in October 2011. Over this 3-month period, the flood wave left heavy extensive damage along its path. While the flooding was massive and the resultant disaster the worst in modern Thai history, the floods did provide an opportunity to review, analyse and assess the planning, response and actions at various stages so that valuable lessons could be learned for improving future flood prevention and water management for the country.

A major policy response is that flooding is no longer considered as just a natural occurrence, but one which is closely linked to human activities and socio-economic developments to meet the food, energy and infrastructure needs of an ever growing and demanding population. Unfortunately, in the haste to develop, the infrastructure designs have sometimes overlooked the impact of natural water flow and runoff, and their effects upon river and canal flows. One key observation of the 2011 floods was that the drainage canal systems in the central plain of Thailand could not handle the huge volume of flood waters. There is also greater recognition that many flood management infrastructures were simply inadequate in their engineering design parameters, outdated construction practices, and lack of regular operation and maintenance procedures towards flood protection. As a response, the government has now begun a huge programme estimated to cost around 100 billion bahts (US$3.3 billion) to repair, rehabilitate and upgrade the flood infrastructure.

In more general terms, the main causes for the recent floods were climate (namely rainfall), management and infrastructure. Therefore, flood management will now focus on approaches to better address these three main causes. Firstly, in the area of climate change, there is now greater awareness that extreme hydrological events will become more frequent and the perceptions, practices and policies of all concerned agencies, including decision makers and the decision-making cycle, should be reviewed and reassessed. To cope with such changing climatic conditions, Thailand needs to develop technologies to predict/forecast more precisely and in advance, the occurrence of rain storms and typhoons, and their intensities and expected impacts, so that there will be sufficient time to take preventive and protective measures and provide early warning to the population.

Secondly, in terms of management, nature can no longer be taken for granted and 'business as usual' is no longer tenable. With flooding becoming an ever greater challenge and the economic and social costs mounting, it is imperative that sound and timely flood management planning be undertaken. The planning process should consider the entire spectrum of upstream to downstream factors and include consultations with all stakeholders who are affected by flooding, with a view to preparing an over-arching flood and water management plan that mitigates, as far as possible, catastrophic flooding such as that witnessed in 2011. Moreover, it is essential that any plan strikes a balance between hard and soft engineering, and also recognizes the critical role that economic and urbanization influences will contribute to flood discharges, and the measures that the government should take to manage in a responsive way.

A policy response being considered is to pay greater attention to land use and watershed management. For decades, the rural population in remote areas have engaged in deforestation activities both for timber as well as to open up new areas for agriculture. In recent years, such practices have increased significantly due to the practice of mono-culture cash crops such as orchard, corn, rubber and palm oil. This problem is further compounded as the destruction of forests in the watershed area has not only increased the runoff leading to more frequent flash floods, but also caused heavy soil erosion and landslides leading to increased siltation of the reservoirs. Post 2011, there have been calls for an urgent national policy review on future agricultural watershed development to more adequately address the watershed land use challenge. At the same time, a comprehensive and more holistic flood water diversion management plan (which uses flood simulation extensively), combined with consultations with local stake holders, is being considered.

Henceforth, the principle of risk management will be the key to reduce the damage from extreme climatic events. Decision makers will need to set priorities with respect to social and economic sectors as well as potential affected areas so as to reduce overall damage. Furthermore, water resources will have to be managed in integration with other natural resources, and the principle of Integrated Water Resource Management should be applied and well understood.

Thirdly, with respect to the water infrastructures, the recent water-related disasters have highlighted the need to review engineering design concepts and design criteria by focussing more on flood protection and flood control. This is because many of the existing structures were designed and constructed based on normal hydrological conditions, while some did not take into consideration hydraulic conditions as well as the livelihoods of people around the affected areas. As such, it will be necessary to review and reset standards in response to climate change which will result in extreme hydrological events with which the old designs cannot cope. In addition, a review of town planning and zoning will also be necessary due to the fact that some communities or even industrial estates were wrongly located in the flood plain, as the planners had overlooked the risk of flooding. For the long term, people will have to learn to live under extreme hydrological events and adjust by building resilience capabilities in their communities.

In summary, living with floods in Thailand is not a new phenomenon and it is likely that the 2011 floods could be a good catalyst to remind government and communities that the time is now right to develop sound flood preparedness and mitigation measures that encompass a balanced use of structural and non-structural measures. Nature possesses powerful forces that are at times difficult to deal with. Perhaps a philosophy of learning to live with nature is an option to consider, whereby future flood management planning and preparedness recognizes and strikes a better balance between nature and engineering.

4.2. The SMART Tunnel: an underground approach to flood mitigation

4.2.1. Introduction. Malaysia is located between latitude 1° and 7° north, and longitude 100° and 119° east. The country comprises two regions, Peninsular Malaysia and the States of Sabah and Sarawak, separated by 640 km of the South China Sea. Peninsular Malaysia adjoins Thailand in the north and stretches down to Singapore in the south. Sabah and Sarawak are situated, respectively, in the north and northwest of the island of Borneo. Together, the two regions cover an area of 330,000 km^2. Located in the humid tropics, Malaysia has an average annual rainfall of 3,000 mm.

The city of Kuala Lumpur, the capital, was founded in the late 19th century as a tin mining settlement and, over the years, has grown to become Malaysia's largest and most important city. Kuala Lumpur is situated at the confluence of two tributaries of the Klang River which originates from the central mountain range at an elevation of 1,330 m and traverses a distance of nearly 120 km before discharging into the Straits of Malacca, draining a catchment of about 1,288 km^2.

The river's headwaters comprise mountainous and steep terrain covered almost entirely by a thick canopy of tropical jungle. The mid upper reaches where Kuala Lumpur is situated is generally less steep and lies between 30 and 60 m above mean sea level. Downstream of the city, the river flows through gently rolling lands and a flat coastal plain before discharging into the sea. The rolling grounds which were earlier opened up for agriculture, have in recent times been developed into new townships and residential areas, and this process of land use conversion is expected to continue into the foreseeable future.

4.2.2. Flooding situation. Topographically, Kuala Lumpur was built along the flood plains of the Klang River and thus, from the earliest days, had been subjected to flooding. The 1926 flood, the so called 'Bah Merah' (Red Flood) is believed to be the worst in living memory but the most notable and extensive flood event on record was the flood of January 1971. This flood lasted for 5 days, cut off the city from the rest of the country, and resulted in extensive damage to property, infrastructure, agricultural land and crops, and in some loss of lives. About 445 ha of land in the city were inundated to depths up to 2 m causing widespread damage and huge losses to infrastructure and property. Since then, the city has experienced a number of major flood incidents (Table 3). A major flood event is defined as one when an area of >100 ha of built up area is submerged to 0.5 m or more.

There are two basic types of floods which occur in the Klang River basin. The first is the monsoonal type flooding caused by a long duration (3–10 days) of low intensity rainfall (>20 mm/h) precipitating over a large area, resulting in the major river systems over-spilling the banks and causing widespread flooding. The flood of 1971 was a monsoon type flooding. Flooding of the second type is caused by thunderstorms which are localized rainfall of very high intensities (>180 mm/h) and short durations (2–5 h). The intense level of precipitation causes large volumes of storm runoff even from small catchments, which often leads to drainage and river systems in the localized areas being completely overloaded. This causes what is termed as 'flash' flooding. As the name suggests, flash floods are quick to manifest and equally swift to subside. This form of flooding is frequently experienced in Kuala Lumpur and it can happen several times in a year.

The frequency of flood events (both monsoon type and flash floods) has increased over the years. The main causes of this can be attributed to the extensive urbanization of the catchment up to the hill slopes, while the capacity of the tributaries are at the same time greatly reduced by the heavy siltation due to the uncontrolled land clearing activities and wanton disposal of solid wastes in the rivers. This disturbance to the eco-balance by converting forests into townships has greatly increased the area of impervious surfaces and consequently reduced the flood detention capacity. Studies in Malaysia have shown that forested catchments capable of absorbing 100 mm of rainfall in the 1st hour of a storm can only absorb 20 mm of rainfall when converted to urban conditions. The increasing trend of the flood magnitudes is best illustrated in the measurement of annual flood discharges taken at Sulaiman Bridge (near the city centre). From the mid-1980s there has been a 300% increase in the mean annual flood discharges from 148 to 440 m^3/s (see Figure 1).

As a result, over the past three decades, the incidence of major flooding has become more frequent (see Table 3). In addition to these major flood events, there have been numerous occurrences of flash floods in the city. These descend with very little warning and totally upset city routine, disrupting commerce and traffic. Overall, the situation is expected to be further aggravated due to the impact of global warming and climate change.

Table 3. Major flood events in Kuala Lumpur (Abdullah, 2007).

Period	No. of times	Year
1970s	1	1971 (often referred to as the 'Great Flood').
1980s	3	1982, 1986, 1988
1990s	4	1993, 1995, 1996, 1997
2000–2010	6	30 April 2000, 26 April 2001, 29 October 2001, 11 June 2002, 10 June 2003, 10 June 2007

4.2.3. Policy responses. In the aftermath of the 1926 flood, the then colonial government initiated steps to resolve the flood issue. The policy response was to 'control' floods through structural (engineering) measures. The early flood control works were carried out by the Hydraulic Branch of the Public Works Department. In 1932 the Department of Irrigation and Drainage (DID) was formed and it took over this function. Some of the early works were:

(i) channelization and protective works of the Klang River, completed in 1933;
(ii) dredging of the Klang River between 1937 and 1941;
(iii) improvement of the Klang River through the city, completed in 1960;
(iv) improvement of the Gombak River, completed in 1969.

Nevertheless, as the level of development in the pre-war years was not significant, there was no major flood in the ensuing two to three decades, though the problem of localized flooding due to inadequate drainage continued to persist. However, as Kuala Lumpur continued to transform into the ultra-modern metropolis it is today and as development kept changing the face of the city, the existing infrastructures came regularly under pressure to cope with flood flows beyond their design limits. This urbanization of the catchments in and around the city has resulted in much upheaval to the hydrological regime and has increased flood runoffs many fold, overstraining the drainage capacity of the river system, and bringing with it challenges for the engineers to come up with innovative engineering solutions.

After the disastrous flood of 1971, the policy response was to entrust hydrology and flood mitigation as additional functions of the DID. The DID then modified the approach to 'mitigate' (reduce the impact) rather than attempt to 'control' floods. This approach involves a combination of structural and non-structural measures. Non-structural or 'soft' measures include flood zoning, flood forecasting and warning, flood risk mapping and flood insurance. Using this approach, the DID developed the Kuala Lumpur Flood Mitigation Project (KLFMP). The main project features consisted of three dams located upstream of the city and urban drainage works on the main tributaries. The project consisted of the following measures:

(i) Channel improvement: this involved the straightening, widening and deepening of 47.2 km of river channel and the removal of the Puchong Drop Structure. Within the city, where space is restricted, the river sections were deepened and concrete lined.
(ii) Klang Gates Dam modification: the height of the existing Klang Gates gravity arch concrete dam completed in 1959 was raised by 3 m to create a new active capacity of 28.8 million m^3, i.e. 76% over the 16.3 million m^3 previous capacity. An additional land area of 71 ha of forest reserve was required for the reservoir impoundment. New spillway piers, gate hoists and four radial gates were installed in the spillway. All the works were completed in 1981.
(iii) Batu Dam and reservoir: the construction of a new earthfill embankment dam with a reservoir capacity of 36.6 million m^3 to control surface runoff from a catchment of 50.2 km^2. This multi-purpose flood control and water supply dam was completed in 1987 and provided an additional 1.34 m^3/s (25 million gallons per day) of domestic and industrial water supply to the city.
(iv) Gombak Dam and reservoir: this was a proposed multiple-arch concrete structure of 25 m height and a reservoir capacity of 42.6 million m^3 covering an area of 82.3 ha. However, as it required an additional 411 ha of land for roads, services and other project functions, there was much

social opposition from the 1,110 families residing in the proposed reservoir area, and the Gombak Dam project was eventually shelved.

(v) Batu Retention Pond: the Batu Retention Pond was proposed as a partial replacement for the Gombak Dam. The Batu Pond has since been completed (in 1993) and the Gombak River Diversion (3.4 km) was completed in 2003. The Pond is designed to carry a flood discharge (100 years return period) of 60 m^3/s from the Gombak River via a diversion channel, and 40 m^3/s from the Batu River.

(vi) Pumped Drainage: Kampung Baru is an inner-city area of about 90 ha located on the right bank of the Klang River about 2 km upstream of the Klang/Gombak confluence. About 35 ha of this area is below the designed level of the Klang River and, of this, 15 ha suffers from flooding due to internal water inundation as the area is relatively low. To mitigate the flooding, the area was 'poldered' and a pump house and a storage basin was constructed alongside the embankment of the river in 1993.

Work on the KLFMP commenced in the mid-1970s, but the pace of implementation was limited by the availability of funding. In the meantime, the rate of development of Kuala Lumpur and the surrounding environs picked up at a much higher level and consequently increases in flood discharges began to outpace the improvement to the river capacity through the flood mitigation works. After the series of floods that hit Kuala Lumpur from 2000, it became apparent that a completely new approach was necessary to alleviate the flood problem in the city. Among the options considered was the possibility of providing additional upstream storage for the excess flows, increasing the capacity of the rivers and creating flood bypasses to divert flood flows away from the city centre.

Detailed studies indicated that for the northern part of the catchment, a number of disused tin mining ponds could be converted into flood detention ponds, while for the southern part, the only viable alternative was to construct a flood bypass. Here, the topography of the land was such that a bypass alignment would cut through high ground as well as areas which were densely built up, making the use of an open channel (open cut) prohibitive in terms of land acquisition and compensation costs. It was thus decided that the bypass should be a tunnel as this was a more competitive alternative in terms of pricing and would offer the least disturbance during construction.

At the same time, it was realized that a flood bypass tunnel would be in use very infrequently and hence, by incorporating a road into the tunnel, the project could be value added by helping to reduce the traffic congestion at the southern gateway of the city to the city centre. Thus was born the idea of the SMART project.

4.2.4. The SMART Tunnel[3] project. The SMART Tunnel project comprised two components; a stormwater component consisting of a tunnel flood bypass to mitigate the flooding problems in Kuala Lumpur city centre and a motorway component using the same tunnel to provide an alternative traffic dispersal scheme to ease the traffic congestion at the southern main gateway into the city. The project was designed to provide Kuala Lumpur with a Q_{100} protection and this required the use of a very large bore tunnel for the bypass.

The stormwater component was designed to reduce the storm flow through the city centre to a manageable quantity within the capacity of the river. A total of 280 m^3/s of flood discharge would be

[3] Although the term SMART Tunnel is a misnomer since the 'T' in 'SMART' already stands for tunnel, this term was the official name of the project to highlight that it was a unique and innovative use of a tunnel.

diverted into a holding pond with an area of 8 ha and having a capacity of 600,000 m^3. The water would then flow into a bypass tunnel 9.8 km long and with an internal diameter measuring 11.8 m. The total storage provided by the tunnel during diversion is 1 million m^3 and it discharges into a storage reservoir of 23 ha, having a capacity of 1.4 million m^3. The effect of the bypass and the storage that can be contained within the system is to reduce both the flow through the city centre as well as the subsequent discharge back into the river system downstream of the city.

The motorway component comprises a 3 km stretch at the middle of the tunnel which offers motorists from the south a speedy alternative to the central business district. To ensure safer traffic operations, this portion of the tunnel was constructed with two decks: one for motorists entering and the other for exiting the city centre. Owing to the limitation of headroom in the double deck configuration, only cars and other light vehicles are allowed to use the tunnel. Figure 6 shows a cross-sectional view through the motorway component.

The SMART Tunnel is operated in three modes based on the relationship between flood discharge and the operation status of the motorway (Figure 7). In mode I (no storm), which is the normal condition, the tunnel is kept dry, as no water is diverted into the system. Twin floodgates are installed at either end of the traffic section to isolate this part of the tunnel from the other sections.

In mode II, (minor storms) some water is diverted into the tunnel but is confined to the lowest drainage chamber provided in the traffic tunnel. During such times, the set of twin floodgates at either end of the traffic tunnel are kept shut, to ensure safety for the traffic in the tunnel. Each gate in the twin set is,

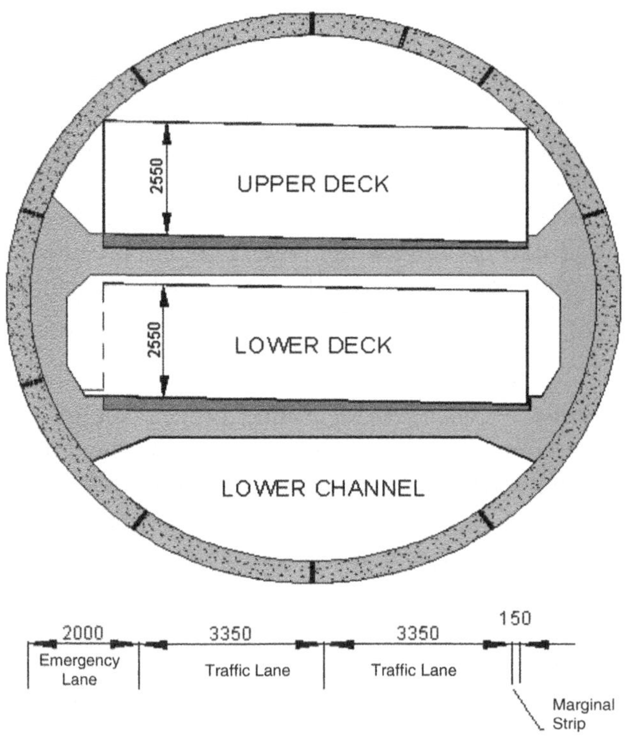

Fig. 6. Typical cross section of motorway tunnel (Abdullah, 2007).

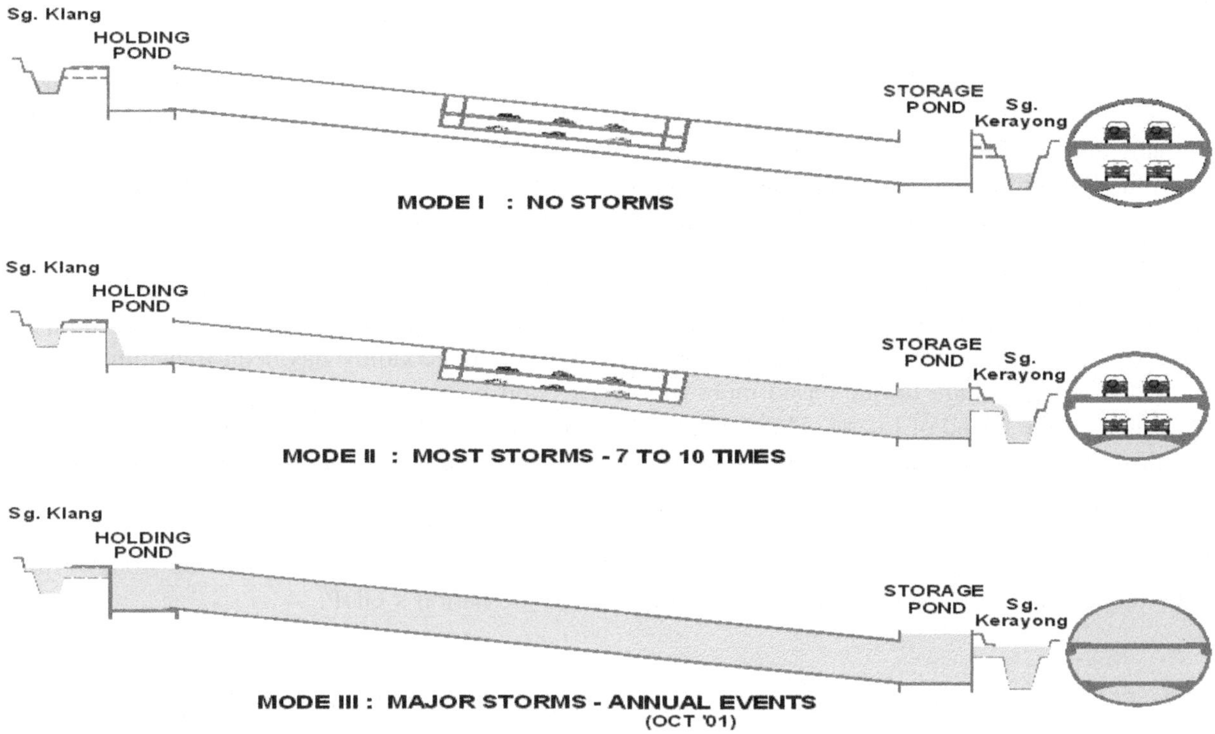

Fig. 7. Operation of SMART Tunnel (Abdullah, 2007).

by design, capable of sealing the traffic compartment. Nevertheless, for additional safety reasons, a second set was provided for backup.

In mode III (major storms), a much larger discharge has to pass through the tunnel, for which the full section of the traffic compartment is required. During this operation, the tunnel is closed to traffic and secured for flooding. Road gates placed at either end of the traffic compartment prevent water in the tunnel from reaching the ground surface at the ingress/egress.

4.2.5. Financing mechanism for the SMART Tunnel project. Traditionally, flood infrastructure is funded by the government social project and there is no attempt to recover cost from the beneficiaries. Capex funding is procured under the 5-year National Development Plans, while O&M funding has to be sourced through the annual operating budget. While there has been private sector involvement in numerous infrastructure projects under the build-operate-transfer (BOT) concept, there have been no takers for flood projects due to the lack of political will to institute a viable cost recovery mechanism, e.g. through local or property tax.

The SMART Tunnel project was funded through a new financing mechanism whereby the capex costs were shared between the government and the private sector, with public funds paying for the flood component and private funds for the motorway component. The private sector was given a concession to recover their costs by charging a toll on motorway users under the BOT concept and was responsible for operating and maintaining the tunnel, thereby relieving the government of its obligations for O&M. However, the government retained control of the operation during floods.

4.3. Super Typhoon Haiyan (Yolanda)

4.3.1. Introduction. The Republic of the Philippines is an archipelago of 7,107 islands located between latitude 4° and 21° north, and longitude 116° and 126° east. To the north lies Taiwan, to the west is Vietnam and to the south and south west sit Indonesia and Malaysia. Including inland bodies of water, the Philippines has a total land area of approximately 300,000 km^2, while its 36,289 km of coastline makes it the country with the fifth longest coastline in the world. Its location on the Pacific Ring of Fire and close to the equator makes the Philippines prone to earthquakes and typhoons.

According to the official count, the population hit 100 million in July 2014, making the Philippines the seventh most populated country in Asia. It is estimated that half of the population resides on the island of Luzon. A newly industrialized country, the Philippine economy has been transitioning from one based on agriculture to one based more on services and manufacturing.

Located on the island of Luzon, Metro Manila, the capital of the Philippines, is a huge urban mass of 16 cities, the most populous of which is Quezon City. Metro Manila is the most populous metropolitan area in the Philippines with a population of over 16 million, while Greater Manila, which includes the suburbs in the adjacent provinces (Bulacan, Cavite, Laguna and Rizal) has a population of around 21 million. Metro Manila's gross regional product is estimated (as of July 2009) to be 468.4 billion pesos (at constant 1985 prices) and accounts for 33% of the nation's GDP.

4.3.2. Flooding situation. Sitting astride the typhoon belt, the Philippine archipelago experiences annual torrential rains and thunderstorms from July to October, with around 20 tropical cyclones entering Filipino waters in a typical year and 8 or 9 making landfall. A tropical cyclone is categorized according to its wind speed. The lowest category is a 'tropical depression' which has wind speeds not exceeding 33 knots (61 km/h). A tropical depression is upgraded to a 'tropical storm' should its sustained wind speed exceed 34 knots (63 km/h); and should the storm intensify further and reach sustained wind speeds of 48 knots (89 km/h), then it is classified as a 'severe tropical storm'. Once the cyclone's maximum sustained winds reach 64 knots (119 km/h) it is designated as a 'typhoon'.

In recent years, tropical cyclones have been increasing in frequency and intensity and as a result, in 2009, typhoons were further divided into three sub-categories: 'typhoon', 'severe typhoon' and 'super typhoon'. A typhoon has wind speeds of 64–79 knots (119–149 km/h), a severe typhoon has winds of at least 80 knots (150 km/h), and a super typhoon has winds of at least 100 knots (190 km/h).

Nearly one-third of the world's tropical cyclones form within the western Pacific with the area just northeast of the Philippines being the most active place. Within the Philippines, the areas most frequently impacted by tropical cyclones are northern and central Luzon and eastern Visayas. These tropical cyclones contribute at least 30% of the annual rainfall in the northern Philippines, while for the southern islands it is under 10%. While the percentage may look small, the precipitation is generally delivered over a few days, resulting in very high rainfall intensities. This rainfall, coupled with the strong winds, invariably culminates in severe flooding. The wettest known tropical cyclone to impact the Philippines archipelago was a 1911 cyclone, which dropped over 1,168 mm of rainfall within a 24-hour period in Baguio.

Typhoons also cause storm surges inland. Historical records between 1897 and 2013 indicate that many typhoons hit the Visayas area near Tacloban City, Leyte, but the five worst storm surges are as shown in Table 4.

Table 4. Historical records of storm surge (*Source:* Nepomuceno, 2014).

Date of typhoon occurrence	Fatalities	Station pressure (hPa)	Storm surge
12 October 1897	1,300	925	7.3 m (Hernani)
24 November 1912	52	924	7 m (Santa Rita)
27 October 1952	444	930	Not recorded
04 November 1984	1,167	925	3.5 m (coastal areas of Leyte)
08 November 2013	6,241 +	910	6–7 m (Leyte & Samar areas)

The vulnerability of the Philippines is best illustrated by two very severe typhoons that have recently struck. In December 2012, Typhoon Bopha (known in the Philippines as Pablo), was the strongest tropical cyclone to ever hit the southern island of Mindanao, making landfall as a super typhoon with winds of 280 km/h. Bopha caused widespread destruction on Mindanao, leaving thousands homeless, causing 1,146 fatalities, with another 834 missing and more than 170,000 people moved to evacuation centres. Total damage was estimated at US$1.04 billion.

Almost a year later, in November 2013, Super Typhoon Haiyan slammed into the Visayas region of the Philippines, killing more than 6,300 people and resulting in US$9.7 billion in losses and damage.

4.3.3. Super Typhoon Haiyan (Yolanda). Super Typhoon Haiyan (known in the Philippines as Yolanda), was the 23rd tropical cyclone to hit the Philippines in 2013 and the 9th that made landfall. Haiyan is believed to have been the strongest typhoon ever to strike land anywhere in the world, and the strongest typhoon ever recorded in terms of wind speed, with wind gusts of up to 200 knots (380 km/h). It is the deadliest Philippine typhoon on record, killing at least 6,300 people in the Philippines alone. More than 2 months later, bodies were still being found.

Haiyan originated from an area of low pressure about 425 km east-southeast of Pohnpei, one of the states in the Federated States of Micronesia, on 2 November 2013. Moving westward, the system encountered environmental conditions favouring tropical cyclogenesis (i.e. the development and strengthening of a tropical cyclone in the atmosphere) and developed into a tropical depression by the following day. After becoming a tropical storm and being assigned the name 'Haiyan' on 4 November, the system began a period of rapid intensification that brought it to typhoon intensity the next day.

On 6 November, the Philippine Atmospheric, Geophysical and Astronomical Services Administration (PAGASA) assigned the storm the local name 'Yolanda' as it approached their area of responsibility (PAR). Over the same day, Haiyan attained super typhoon status as the eye of the typhoon passed over the island of Kayangel in Palau with sustained winds of up to 230 km/h. Prior to landfall in the central Philippines, Haiyan reached its maximum intensity on 7 November in the early evening, with 1-minute maximum sustained wind speeds of 314 km/h, unofficially making Haiyan the strongest tropical cyclone ever observed based on wind speed.

Haiyan entered the Philippine PAR at midnight of 6 November and made its first out of six landfalls on 8 November (4:40 a.m.) in Guiuan, Samar. At landfall, it had maximum sustained winds of 235 km/h with gusts of 275 km/h near the centre. From there, it moved west-northwest crossing Northern Leyte at 7:00 a.m., then to Northern Cebu (9:40 a.m.), Northern Panay (12:00 noon) and Busuanga in Northern Palawan (8:00 p.m.). It exited the PAR on the following day, 9 November at 1:40 p.m. Owing to its massive destruction, the Philippine Government declared a National State of Emergency on 11 November 2013.

Haiyan also brought storm surges which were recorded in several locations along the eastern coastline. According to the 8 November situational report of the United Nations Office for the Coordination of Humanitarian Affairs (OCHA), the storm surge created 5–6 m waves that wiped out most infrastructure, health facilities, schools, basic public services, homes and commercial buildings in several coastal towns/barangays in the islands of Samar and Leyte. Based on a PAGASA-DOST[4] survey of storm surge heights, Tacloban City in Leyte (population 221,000 people), bore the most severe damage, with a storm surge of 5–6 m that also resulted in inundation of 600–800 m and which smashed through coastal communities. From Guiuan to Hernani in Samar, a storm surge was recorded at 6–7 m high with 800–1,000 m inundation.

Super Typhoon Haiyan brought so much destruction, particularly in central Philippines. The affected areas consisted of 12,139 barangays in 591 municipalities and 57 cities in 44 provinces, and more than 16 million people were affected, as shown in Table 5.

Some 70–80% of the houses on the island of Leyte were destroyed, with the low-lying areas of Tacloban City the worst affected. The terminal buildings of Tacloban Airport were destroyed, along with almost all of Tacloban's infrastructure. Over 20,000 houses in the city were damaged, with a large part totally destroyed. Ships were washed inland, cars piled up and trees uprooted. Overall, along the 100 km path of Haiyan, some 1,140,332 houses were damaged, 550,928 totally and 589,404 partially. Most of the damage was caused by storm surges, strong winds and heavy rains that resulted in loss of life, property and infrastructure. Based on the results of a joint Asian Development Bank and OCHA rapid assessment of Tacloban and other locations, and on the data provided by the National Disaster Risk Reduction and Management Council (NDRRMC), more than half of the estimated damage and loss was private property.

The total direct losses in the Philippines are estimated to have reached US$9.7 billion and even though on the global scale these losses appear low (e.g. Hurricane Katrina caused an estimated US$125 billion of direct losses), they made a deep dent in the Philippine economy. Property valued at around 4% of the country's GDP was destroyed. Only about 7% of the losses were insured and the balance cannot be recovered without placing a huge additional burden on the country's population or the national budget. The disaster brought about by Haiyan also illustrated well the large loss amplification factor, whereby a big disaster can result in secondary catastrophes such as prolonged periods of non-accessibility to affected places due to infrastructure destruction, and a strong regional economic downturn leading to a population drain and migration outflow from the disaster zone.

Table 5. Humanitarian costs (as of 3 April 2014) (*Source:* NDRRMC, 2014).

Population affected	3,434,593 families (16,078,181 persons)
Population displaced	890,895 families (4,095,280 persons)
Number of reported deaths	6,293
Number of reported injuries	28,689
Number of missing persons	1,061

[4] A term frequently used as PAGASA is an agency under the Department of Science and Technology (DOST) of the Philippines.

4.3.4. Policy responses. The Philippines is a country where massive natural disasters including typhoons, earthquakes and volcanic eruptions are a way of life. Sitting astride the typhoon belt and having a long coastline of 36,289 km, the Philippines archipelago is particularly vulnerable to tropical storms and typhoons. On average, up to 20 tropical storms enter Filipino waters annually, with about half of them making landfall. The most frequently impacted areas are northern and central Luzon and the eastern Visayas. The capital, Metro Manila, is located on the island of Luzon.

Typhoons bring heavy rainfall which, when combined with the strong winds and storm surges, can cause large-scale flooding and result in enormous destruction and loss of life. In 2013, the Philippines was struck by Super Typhoon Haiyan, the strongest typhoon ever to strike land anywhere in the world, with wind gusts of up to 200 knots (380 km/h) and creating a storm surge 6 m high and extending 1 km inland.

In the face of such daunting climatic factors, it would be prohibitively costly for the Philippines to invest in the necessary defensive infrastructure to protect the country or even to try to mitigate the impacts from typhoons. Instead, the policy response has been to focus on improving disaster preparedness and management. In 2010, the Philippine Government enacted Republic Act No. 101211 on Disaster Risk Reduction and Management (DRRM Act 2010). This Act provided for the development of policies and plans and the implementation of actions and measures pertaining to all aspects of DRRM, including: (i) good governance; (ii) risk assessment and early warning; (iii) knowledge building and awareness raising; (iv) reducing underlying risk factors; and (v) preparedness for effective response and early recovery.

Significantly, the law enabled the government to review and finalize its landmark plan on DRRM 'Strengthening Disaster Risk Reduction in the Philippines: Strategic National Action Plan (SNAP) 2009–2019' based on its global commitment to disaster risk reduction (DRR), as embodied in the Hyogo Framework for Action. The SNAP has been institutionalized through Executive Order No. 888 signed on 7 June 2010. The SNAP recognizes the paradigm shift from a mostly reactive disaster response approach to a proactive DRR orientation.

Further, Administrative Order No. 1 directed all government units, particularly provinces, to adopt (in their planning activities) the guidelines on mainstreaming disaster risk reduction, in sub-national development and land use/physical planning. As the Philippines is considered one of the countries most prone to events triggered by natural hazards, the country's vulnerability to such hazards will continue to hinder socio-economic development unless practical solutions are found to avert potential damage from them. The Medium-Term Philippine Development Plan has identified disaster mitigation as a priority thrust, and achievement of such thrust includes the integration of a disaster management strategy in the development planning process at all levels.

The DRRM Act called for the development of a National DRRM Framework which shall provide for a comprehensive, integrated, multi-sectoral and inter-agency and community-based approach to DRR. The framework shows that mitigating the potential impacts of existing disaster and climate risks, preventing hazards and small emergencies from becoming disasters, and being prepared for disasters, will substantially reduce loss of life and damage to social, economic and environmental assets. It also points out the need for effective and coordinated humanitarian assistance and disaster response to save and protect the more vulnerable groups during and immediately after a disaster. Further, building back better after a disaster will lead to sustainable development after the recovery and reconstruction. To ensure its relevance to the times, the DRRM Framework will be reviewed every 5 years, or as may be deemed necessary.

Inherent in the DRRM Act of 2010 is the transformation of the National Disaster Coordinating Council into the NDRRMC. This Council is made up of 33 representatives from various government agencies, and seven representatives from non-government, civil and private sector organizations. It is chaired by the Secretary of the Department of National Defence and administered by the Office of Civil Defence under the Department of National Defence. The Council is responsible for ensuring the protection and welfare of the people during disasters or emergencies within the framework of DRRM as expounded under the law.

In the aftermath of Super Typhoon Haiyan, disaster preparedness and management has justifiably received major government attention, with funding increased by nearly 100 billion Pesos, or about 5% of the total national budget. However, while a portion has been spent wisely on more proactive and innovative disaster mitigation programmes such as hazard mapping, the bulk of the funds is earmarked for cleaning up, repairs and rehabilitation of the damage from Haiyan.

What would happen if such a super typhoon were to strike Metro Manila? Simulation studies indicate that at least 4,000 people would be killed and 13,000 would be injured. Nearly all commercial and residential structures would experience some degree of damage and total estimated damage would probably exceed US$3 billion. Power generation systems and distribution lines would be downed, leaving millions of homes without electricity for long periods. All telecommunications systems would experience extensive outages. Transportation infrastructure damage would cause major disruptions. The vulnerability of a large portion of the city population, particularly the urban poor, would result in a humanitarian crisis. Not surprisingly, Metro Manila is ranked by the USA as the third most vulnerable metropolitan area on earth with the second most number of people at risk. There are also serious concerns about what such a storm would do to the national economy, as the city accounts for 40% of the country's gross national product.

5. Conclusion

This is a tale of three capital cities in South East Asia. Although all three cities are situated in the same climatic zone, due to their geographical locations, water disasters impact differently on them, whilst policy responses also differ.

Bangkok is located in the estuary of the Chao Phraya River, which has a drainage area of 160,400 km^2 (or almost half the total land area of Malaysia or the Philippines). The policy response to the flooding problem was to 'control' the floods through upstream storage. However, in the light of the unprecedented 2011 floods, the policy response has been modified to learning to live under extreme hydrological events and to adjust by building resilience capabilities in the communities and flood infrastructure.

Kuala Lumpur is situated at the middle reach of the Klang River, a comparatively small but steep river. In the early days, the policy response was to 'control' floods through structural measures and regular investments in flood infrastructure were made. However, the continuing urbanization of the catchments in and around the city has resulted in flood runoffs increasing many fold, bringing with it challenges for the flood planners to come up with more innovative solutions, from both the engineering as well as the financial perspectives.

Metro Manila sits astride the typhoon belt and is particularly vulnerable to tropical storms and typhoons. Typhoons bring heavy rainfall which, when combined with strong winds and storm surges, can cause large-scale flooding and result in enormous destruction and loss of life. In the face of such daunting factors, it would be prohibitively costly to invest in the necessary defensive infrastructure to protect Metro Manila from typhoons. Instead, the policy response has been to focus on improving disaster preparedness and management.

5.1. Policy conclusions

Climate change is already impacting on the hydrological cycle, leading to increased frequencies and intensities of extreme weather events, i.e. floods, typhoons and droughts are occurring more frequently, and are larger and more intense and last longer. In recent years, three large-scale water disasters have occurred in South East Asia. In 2011, the worst floods in Thailand occurred from August to November, and the Philippines was hit by two of the worst typhoons ever recorded: Typhoon Bopha in December 2012 and Super Typhoon Haiyan in November 2013.

Structural measures are still needed to mitigate the impact of such extreme events but the existing water infrastructure will not be able to handle storms which are likely to be several times larger than designed for. Concerted action is needed to repair, rehabilitate and upgrade the flood infrastructure and make it more adaptable and flexible. The design process will have to consider the entire spectrum of upstream to downstream factors, include a balanced use of structural and non-structural measures and involve consultations with all stakeholders to produce an over-arching flood and water management plan that will mitigate, as far as possible, the worst case scenario.

Zero risk does not exist (i.e. engineering works cannot offer complete protection) and in planning for the future, one should expect the unexpected. In terms of management, nature can no longer be taken for granted and 'business as usual' is no longer tenable. Investments in DRR, preparation and prevention will pay off in terms of reduced loss of life, avoided damage, and long-term economic growth and stability.

References

Abdullah, K. (2006). Stormwater Management and Road Tunnel (SMART): An Under-ground Approach to Mitigating Flash Floods. In: *Proceedings of the International Workshop on Flash in Urban Areas and Risk Management, Muscat, Oman, 4–6 September 2006.*

Abdullah, K. (2007). Stormwater Management and Road Tunnel (SMART): A Unique Flood and Motorway Solution in the City of Kuala Lumpur. In: *Proceedings of the 7th Malaysian Road Conference, Kuala Lumpur, 17–19 July 2007.*

Anukularmphai, A. (2014). *Flood Management: Lesson Learned from Thailand 2011 Worst Flood.* Unpublished.

National Disaster Risk Reduction and Management Council (NDRRMC) (2014). Effects of Typhoon 'Yolanda' (Haiyan). *NDRRMC Site Report No. 108*, 3 April 2014. NDRRMC, The Philippines.

Nepomuceno, D. (2014). *Super Typhoon Yolanda, Its Impacts and Aftermath: A Case Study on Policy Actions, Successes and Lessons Learned.* Unpublished.

Translating policies into actions: the case of the Elbe River

Wolfgang Grabs[a] and Hans Moser[b]

[a]*Corresponding author. Federal Institute of Hydrology, Koblenz, Germany. E-mail: grabs@bafg.de*
[b]*International Commission for the Hydrology of the Rhine Basin, Lelystad, The Netherlands*

Abstract

This paper describes methods and processes to link policy development to the implementation of those policies in actionable implementation plans. It is shown that policies can only be implemented effectively if they are embedded in a legal framework that is designed to facilitate achievement of the policy objectives. The paper shows different levels of policy making and decision support for the development of policies at different levels, ranging from the level of Federal States in Germany to policy development and implementation at the European level as part of the European Framework Directive. Using the Elbe River as a case study, the paper shows the need to anchor regional, transboundary and state level policies to mandated national institutions. A key lesson learnt from the Elbe River Basin is that policy integration is of utmost importance. The paper also demonstrates that a balance needs to be reached with regard to structural and nonstructural measures in flood risk management to arrive at a truly integrated flood risk management strategy and its implementation. The development of research policies on the basis of sound science is indispensable in support of policy development and its implementation.

Keywords: Cross-sectorial policy integration; Integrated flood risk management; Policy development and implementation

Introduction

The focus of this paper is on the German part of the Elbe River Basin. Looking at the entirety and complexity of the programs and measures taken to improve flood protection in the Elbe River Basin, the word 'policy' does not appear in any of the references, although the measures described in the IKSE (Internationale Kommission zum Schutz der Elbe; International Commission for the Protection of the Elbe River (ICPER)) Elbe Flood Protection Action Plan (IKSE, 2003) have all the characteristics of an integrated flood risk management plan. This plan is built on policies such as 'giving room for the river', 'living with floods', 'prioritizing long-term spatial management' as means for the retention of floods and lowering the peak of flood waves. To prevent confusion of the use and meaning of the

doi: 10.2166/wp.2015.006

word 'policy', the following paragraphs provide a brief overview. The use of the word 'policy' has several dimensions and is not an exactly defined term but is used in different contexts. Webster's dictionary has a number of closely related definitions. They include: 'A definite course or method of action selected from among alternatives and in light of given conditions to guide and determine present and future decisions'; 'A high-level overall plan embracing the general goals and acceptable procedures especially of a governmental body'. The Oxford Dictionary defines policy as 'a course or principle of action adopted or proposed by an organization or individual'.

Policy objectives reflect the goals that are meant to be achieved, such as minimizing the loss of lives and livelihoods including infrastructure through improved integrated flood risk management. Policy instruments are the means by which policy objectives are implemented through specific actions (ILRI, 1995). Different terms are used to characterize the increasing specificity with which policies are put into practice. This can be viewed as a hierarchy starting from policy formulation through the setting of policy objectives, moving further down this hierarchy to policy-derived 'plans', 'programs' and 'projects'. The formulation of policies requires an in-depth analysis of event-driven identified weaknesses and strengths in flood risk management. The causative factors of weaknesses and strengths, and derived policy options in response to these identified weaknesses and strengths, need to be identified. The identification of policy options is essential as each policy will have repercussions on other policies or ongoing action plans. The development of different flood scenarios has also proved to be a highly useful tool in policy development as these scenarios may project different futures for which appropriate policies need to be put in place. Ultimately, the policy development process leads to a prioritization of policies, as the implementation of policies is closely related to the availability of priority-driven allocation of scarce resources. This last step is usually undertaken at the level of politicians who also make decisions on the applicability of policies.

Policies can only be implemented effectively if they lead to the development of laws, rules and regulations that are designed to facilitate achievement of the policy objectives. Legal frameworks need adequate instruments for their enforcement and, ultimately, policies and legal frameworks can set minimum standards or benchmarks against which improvements can be measured.

As many of the sources cited in this paper are of German origin, it is necessary to note that in the German language there is no direct equivalent of the English word and meaning of 'policy'. In German, 'policy describes the entirety of a political theme area' such as 'Water Politics' (Wikipedia; see: http://www.de.wikipedia.org/wiki/Policy; see also http://www.bpb.de/nachschlagen/lexika/politiklexikon/18014/policy).

A large number of policy contexts exist that drive decision-making in flood risk management. Such policy contexts have been analyzed for England and Scotland with regard to their influence on flood risk management practices with a focus on finding a balanced approach between structural and nonstructural measures (CRUE, 2008). A few policy contexts are mentioned here as examples: policies regarding land use, housing, agriculture, industry, ecology, insurance policy and climate change. It is evident that all these policies, that have probably in most cases been developed in isolation from each other, influence to a varying degree the development of integrated flood risk management policies on the basis of an entire river basin. The comparative influence of different policies on decision-making in flood risk management (including finding a balance between structural and nonstructural measures) is difficult in the absence of metrics to quantify policy influences. The authors of the CRUE (2008) report state, for example, that it is not possible to quantify the impact of a policy on climate adaptation on flood risk management over the next 20 years.

Being already difficult in more centrally governed countries such as the United Kingdom, the problem is aggravated in countries that have a clear federal structure, such as in Germany. In the Elbe River Basin, policy contexts at different government levels are manifest: federal policies, the sector policies of 10 federal states, policies of riparian countries (especially those of the Czech Republic) that have a strong influence on flood risk management in the German part of the Elbe River, and policies at the level of the European Community. It needs to be noted that most of these sector policies have their own legislation and governing administrative super-structures ensuring adherence to and implementation of these different policies.

The key approach under these circumstances is that of 'policy integration'. The main thrust of policy integration is to adapt existing policies to be non-contradictory with respect to flood risk management. The baseline condition is a general acceptance by politicians, interest groups and the general public that integrated flood risk management (including all relevant sector policies) has a higher societal benefit than benefits derived from single-sector driven policies.

Description of floods in the Elbe River Basin

The Elbe River is the fourth largest river basin in Central Europe (Figure 1) with a length of 1,165 km and a catchment area of 148,000 km² of which the majority is in Germany (97,000 km²) and in the Czech Republic (50,000 km²), in addition to smaller parts of the river basin amounting to 1,000 km² located in Austria and Poland (Figure 2). The population of the basin is about 25 million. In terms of flood management it needs to be noted that the hilly character of the basin in the Czech Republic results in short flood forecasting and warning times because of the short time interval between floods occurring and the antecedent rainfall event. Furthermore, flood characteristics in Germany are always

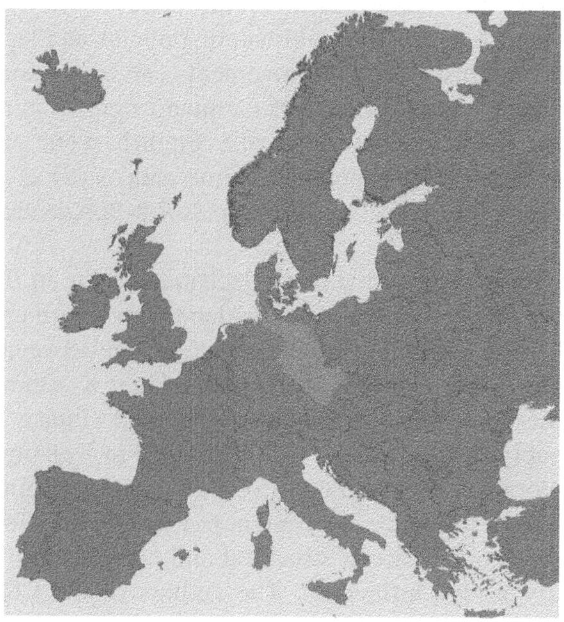

Fig. 1. Location of the Elbe River Basin in Germany (Federal Institute of Hydrology, Koblenz, Germany).

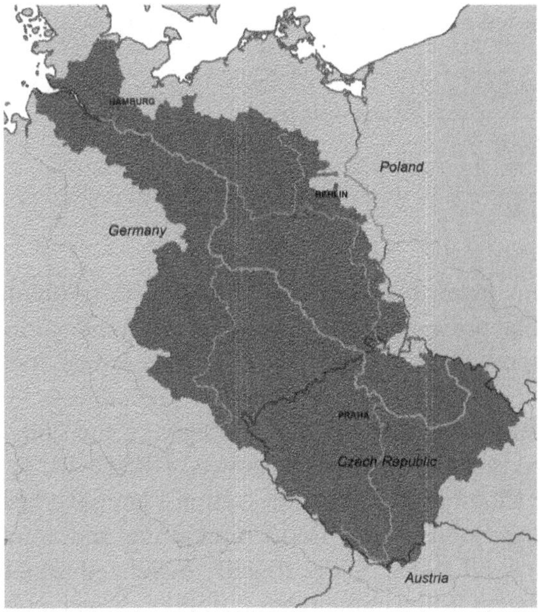

Fig. 2. Map of the Elbe River Basin (Federal Institute of Hydrology, Koblenz, Germany).

dependent on upstream flood management in the Czech Republic with its numerous dams and reservoirs (Raadgever, 2005).

The Elbe floods of 2002 exceeded total costs of just above 20 billion[1] euros in the Elbe River Basin, with some 13 billion euros in Germany alone and about 7 billion in the riparian countries of the Elbe River. These damages amounted to the highest-ever damages from floods in the region at that time (Plate, 2011).

Half of the damage occurring in Germany impacted on public property. About 7.8 billion euros of tax-payers money was used for immediate assistance to flood victims in Germany (WWF, 2007). In 2013, the total damages in Germany amounted to an estimated 12 billion euros in comparison to an estimated 17 billion euros of flood damage in Europe (Zurich Insurance, 2014a).

The overall 2013 flood damage costs in the Elbe River Basin (German part) were considerably lower compared to the 2002 flood loss costs (see Table 1). Although in the Elbe River Basin the number of insured damage cases rose from 150,000 in 2002 to 180,000 in 2013, the Association of German Insurance Companies (GDV) stated that individual damages were lower than in 2002 (Cash Online – Finanznachrichten, 23 July 2013). This could be attributed to some extent to precautionary measures by the affected population which in this context underlines the positive effects of flood risk awareness campaigns that started after the 2002 flood event. It is also important to note that the damages incurred during the 2002 floods were partially attributed to flash floods occurring in the Erzgebirge. In the 2013 flood event, flash floods with their highly destructive character did not occur.

After the disastrous 100-year flood event in 2002 and more local flood events in 2006, 2010 and 2012, another extreme flood event with a statistical recurrence period of over 100 years occurred in

[1] Billion $= 10^9$.

Table 1. Gross comparison of flood losses for the Elbe floods in 2002 and 2013 (adapted from Belz (2013) and LAWA (2014a, b).

Elbe floods	Material damage	Fatalities	Direct federal flood aid
2002	11,710	38	9,000
2013	5,200	15	8,000

All monetary amounts in million euros.

the Elbe River in 2013. Lessons learnt from the 2002 flood event resulted in an efficient flood warning and an information system that prevented the loss of lives (Bundesministerium des Inneren, 2013). However, despite partially improved lead times for flood forecasting, in-time evacuation and a largely improved flood protection, the overall recorded damage to buildings, infrastructure and livelihoods (farms and production sites) surpassed the reported damages of the 2002 flood event.

An analysis of the 2013 floods in the Elbe River showed (IKSE, 2013) that the measures undertaken in the framework of the IKSE Elbe Flood Protection Action Plan have been largely effective, including flood retention measures in the flood plains, flood forecasting and warning, flood preparedness and improved information systems at all levels, with officially mandated organizations and volunteer organizations, and with regard to successfully implemented flood risk awareness programs.

Policy actions taken or not taken

As policy paradigms change from 'flood protection' to 'flood risk management', it is important to recognize that flood risk management should not just aim to protect people from floods but should go much further, aiming to increase the resilience of potentially affected people–the ability to cope with floods and to recover from flooding. Another dimension of risk management is knowledge of possible cascading failures, meaning that if one structure fails, it could cascade into multiple failures (such as the flooding of a power transformer station that leads to the failure of water pumps, which in turn results in further damage to buildings and property) (Zurich Insurance, 2014b). Forward looking policies in flood risk management always carry a measure of uncertainty. The additional adoption of a policy of 'no regrets and flexibility' ensures benefits even when the actual hazard and associated risk is not quantitatively known for a specific time horizon. It also needs to be noted that risk management consists of two major elements: the factual risk analysis that is solely based on physical facts, and the assessment of risk, which is a political process that includes socio-economic values and the acceptance of risk (Deutscher Bundestag, 2013).

Transboundary policy actions: organizational level

Several rivers in Central Europe are transboundary, including the Rhine and Elbe rivers. As a consequence of the large floods in 1988 (CHR-KHR, 1990), 1993 and 1995 (BfG, 1994; CHR-KHR, 1999), the 'Rhine Action Plan' that had been elaborated by the International Commission for the Protection of the Rhine River (ICPR) was adopted in January 1998 (IKSR, 1998). On the level of the Elbe River Basin, the IKSE Agreement was signed in October 1990, with signatory countries being the Federal Republic of Germany and the Czech Republic. Observers are Austria, Poland, European Union, International Commissions for the protection of the Rhine River, the Odra River, and the Donau River and non-governmental organizations (NGOs). Flood protection has become one of the focal areas of activities of the IKSE (http://www.ikse-mkol.org). Triggered by the Elbe floods in 2002, the IKSE developed and commissioned its Elbe Flood Protection Action Plan in 2003. As with the Rhine Action Plan, this plan is not legally binding. It rather serves as a set of recommendations. However, as the participating organizations are mainly governmental organizations of the signatory states, the plan has served as a basic policy document that was largely drawn in the political decision-making process aiming to implement the plan.

National policy actions: organizational level

In Germany, federal waterways (basically the main stems of all navigable large rivers in Germany) belong to the Federal Government. Flood protection is a responsibility of the German federal states ('Länder'). In the light of the 2002 floods in the Elbe, the 10 'Länder' that are situated in the Elbe River Basin founded the Flussgebietsgemeinschaft–Elbe (FGG–Elbe; Elbe River Basin Community) in 2004 (Pressespiegel, 2014). Since the FGG-Elbe's new Administrative Agreement in 2010, it has been coordinating the implementation of both the EU Water Framework Directive and the EU Flood Risk Management Directive (http://www.fgg-elbe.de/start-en.html). As cited in http://www.lawa.de/About-LAWA.html, the federal states also work with each other in a cooperative framework (Arbeitsgemeinschaft (ARGE) Elbe) and with the German Working Group on water issues of the Federal States and the Federal Government, represented by the Federal Environment Ministry (LAWA). Due to the number of actors involved in decision-making, the decision-making process can be lengthy, based on aiming to achieve best possible compromises and balancing the interests of different groups. The strength of policy making in the German context is 'the obligatory and widespread hearing of experts and interested parties of all relevant groups and in the democratic and constitutional consideration of various interests and viewpoints' (Rudolph & Block, 2001).

Decision-making at national level

In the German context, decision-making is based on commonly agreed objectives in a political framework; this could be termed as 'policy'. Integrated flood risk management includes the integration of different data sources in terms of hydrologic, topographic, hydraulic and socio-economic domains, aggregated in a modeling framework. Policy planning to improve integrated flood risk management is therefore highly complex as different scientific domains and the integration of multiple sources of data need to be considered. To give an example: The VERIS – Elbe project, which aims at the modeling of change and management of risks of extreme floods in large river basins (Figure 3), looks at two dimensions of integration (VERIS – Elbe, 2013; Burek & Rademacher, 2009):

> *'Disciplinary models are coupled in a model system of the flood risk system. Results of the model simulations are put in a planning context that ranges from the ex-post analysis over the formulation of scenarios including development trends and mitigation options to their ex-ante analysis, multi-criteria assessment and implementation.'*

Flood risk is calculated as a function of probability × vulnerability on the basis of a series of maps (Figure 4) that allow the quantification of flood hazards and potential damage per unit area (or specific

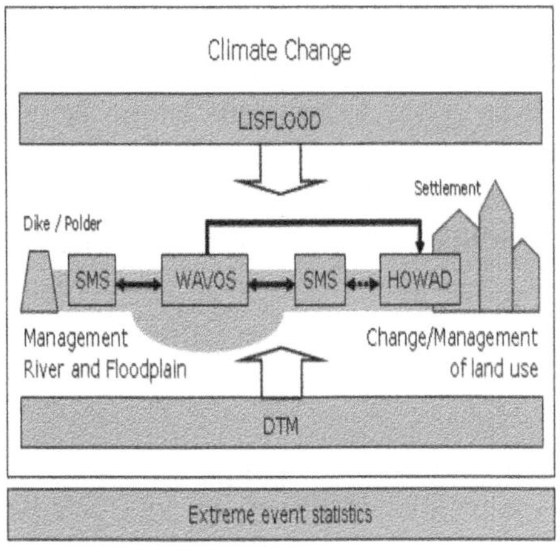

Explanations:

LISFLOOD: Rainfall-Runoff Model;
SMS: Surface Water Modeling System
(2D-hydrodynamic-numerical modeling system for surface waters);
WAVOS: Water Level Prediction System
HOWAD: Flood Damage Simulation Model
DTM: Digital Terrain Model

Fig. 3. Schematic view of coupled models.

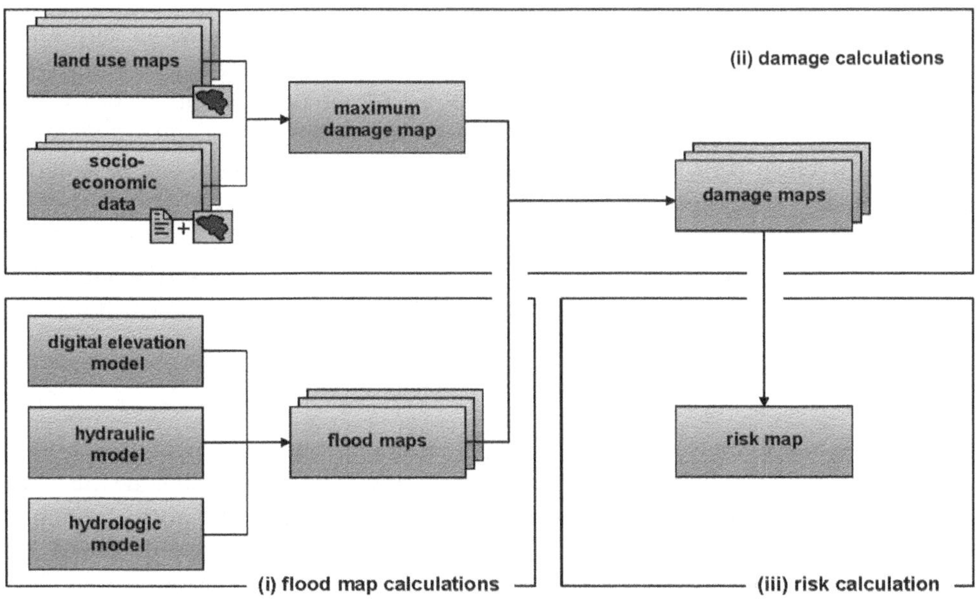

Fig. 4. Schematic view of the development of flood maps and calculation of flood risks.

infrastructure). The calculation of damage and risk consists of three steps, namely: (i) defining probability and extent of flooding; (ii) determining expected damage; and (iii) defining risk. Figure 4, taken from (Vanneuville *et al.*, n.d.), demonstrates this approach.

Under the impact of the 2013 flood events in Germany (BfG, 2013) the environment ministers at Federal and State government levels agreed in June 2013 on the development of a national flood protection program. The national flood protection program was passed by the German conference of state environment ministers in October 2014. The two-pronged approach envisages the amendment and change of relevant administrative procedures with the objective of simplifying and accelerating the planning and implementation of measures with a focus on preventive flood protection. Federal Government and the State governments see it as an obligation to initialize preventive investments in a national flood protection program.

The Extraordinary Conference on Floods of the Environment Ministers of the German Federal States in September 2013 stated that the recent flood events might be early signs of climate change and decided that a national flood protection program would be developed including:

- review and eventual further development of design-flood and mutually coordinated approaches to assess impacts of potential flood protection measures;
- prioritization of preventive flood protection measures with a focus on supra-regional measures and on the basis of additional flood retention areas that will have a significant effect on the lowering of peak flood levels;
- elimination of deficiencies in existing flood protection measures;
- development of a joint (Federal and State level) financing strategy.

The ministerial conference further decided that as a matter of policy, the national flood protection program should be based on the following concepts that will be underpinned by adequate measures and projects:

- making room for rivers through the backwards re-location of dykes and the establishment of additional polders allowing controlled flooding;
- maintaining and re-establishing the function of flood plains in the future through long-term spatial management;
- this also includes reviewing design-flood criteria.

Recognizing further that effective flood protection in the general interest cannot be compromised by singular interests, in July 2013 Saxony together with Bavaria brought forward a draft law on the acceleration of flood protection measures (Hochwasserbeschleunigungsgesetz) (Wiederaufbaustab Sachsen, 2013). In its essence, the law will lead to an accelerated administrative and legal process to plan and implement flood protection measures without lengthy administrative procedures.

At the highest national political level, the national flood protection program has been taken into the Coalition Contract of the Federal Government for its 18th legislation period stating that, under the overall coordination of the Federal Government, a national flood protection program will be elaborated in cooperation with the Federal States. As stated above, the thrust of the program aims at the development of supra-regional measures for preventive flood protection as well as the standardization of flood protection criteria in German river basins (Koalitionsvertrag, 2014).

The coalition contract also states that with respect to flood management the Elbe integrated river basin management plan will be implemented with a balance between ecological and economic requirements. Two other policy elements are important to mention as these are indispensable in integrated river basin management in general and integrated flood risk management in particular:

- entering into an intensified dialog with riparian countries on flood protection; and
- considering the repercussions of integrated flood management practices on other sectors (including agriculture, public and private infrastructure) and their own administrative and legislative frameworks;

it is envisaged that a federal spatial management plan with common standards regarding flood hazard areas, polders, etc., will be developed.

Policy at the level of the European Community

Building on the experiences of the Rhine and Elbe action plans and strongly advocated by the Federal Government of Germany together with riparian countries, the European Commission put in force a Directive on the assessment and management of flood risks (EC, 2007).

This Directive builds on policies on flood risk management that were communicated inter alia to the European Parliament in 2004 (EC, 2004). This communication sets out the analysis and approach to managing flood risks at the level of the European Community. The Directive also makes reference to Directive 2000/60/EC (EC, 2000) that established Community action in the field of water policy, requiring river basin management plans. These two Directives (EC, 2000, 2007) are the guiding directives for Integrated River Basin Management in the overall context of Integrated Water Resources Management (IWRM).

The introductory parts of the Directive (EC, 2007) provide a policy framework including:

- space for rivers;
- components of flood risk management plans;
- use of 'best practice' and 'best available technologies';
- coordination with riparian countries in transboundary river basins.

Within this policy framework, sub-policies have been adopted by the Commission including more detailed instructions in the fields of:

- Preliminary Flood Risk Assessment;
- Flood Hazard and Flood Risk Maps;
- Flood Risk Management Plans and policy directives outlining;
- Implementation measures.

It needs to be noted that national and state flood policies and action plans (including the European Flood Action Programme (2014)) and consequently the EC Directives were all triggered by disastrous flood events; the EC (2004) Communication on Flood Risk Management cites as the starting point of its inception that 'Between 1998 and 2002, Europe suffered over 100 major damaging floods, including the catastrophic floods along the Danube and Elbe rivers in 2002.'

Risk governance

The concept of risk governance is important as it involves the rational assessment of risks and adequate administrative, institutional and legislative arrangements. In the case of the European Flood Directive, the initiated planning process resembles the risk governance concept as described by the International Risk Governance Council (Heintz *et al.*, 2012).

Risk governance can consist of the four steps shown in Figure 5 (Heintz *et al.*, 2012); risk evaluation and risk management actions are then documented in Flood Risk Management (FRM) plans.

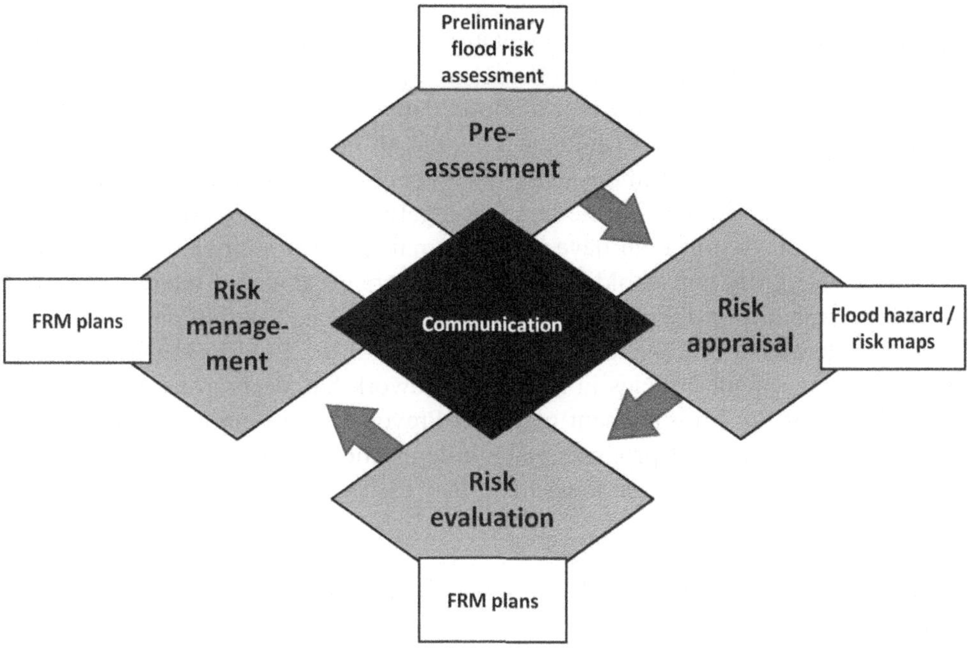

Fig. 5. Risk governance concept in flood management (Heintz *et al.*, 2012).

Communication is the central element of risk governance with a focus on stakeholder participation at the lowest appropriate level.

Successes and inadequacies

Successes

An ex-post analysis of the 2013 floods in comparison to the 2002 floods showed (Rickmeyer, 2014) that, based on findings of the Federal Institute of Hydrology, Koblenz, Germany:

- Peak water levels of the main flood wave in the German part of the Elbe River would have been 20–75 cm above the observed levels, if intended and unintended flood retention actions had not been in effect.
- Breaches of dykes and the controlled flooding of parts of the flood plains lowered the highest water levels at the middle Elbe River between 34 and 46 cm.
- If the reservoirs in the Czech Republic and in the German state of Thuringia had had a higher retention potential at that time, water levels could have been lowered between 20 and 66 cm along the German part of the Elbe River.

Many of the actions envisaged in the IKSE Elbe Flood Protection Action Plan have been implemented since 2003. However, if all measures planned had been implemented by the time of the 2013 flood, an additional lowering of the peak water levels in the middle Elbe could have been realized in the order of 20–40 cm.

IKSE states (IKSE, 2012a, b) that the Elbe Flood Protection Action Plan has been validated as a useful instrument in flood risk management. The implementation included mainly structural and to a lesser degree nonstructural measures in the policy framework of the Action Plan. The subsequently implemented measures have resulted in an overall reduction of flood damages and losses and a significant reduction in direct Federal aid for flood victims (Table 1). This has to be seen against the background that the floods in 2013 were even more severe in the sense that the entire Elbe River Flood Basin was affected.

The overall success story is also due to the fact that legally non-binding policies such as those of the IKSE Elbe Flood Protection Action Plan have been internalized in a political decision-making process with inputs from all levels, from household level upwards through a host of interest groups. The second positive factor has been that all political measures were fully backed within an existing (or amended) legal framework at Federal and State level in concurrence with the EC Flood Directive.

Embedding flood management policies in a legal framework has been an asset in policy implementation. The Federal Act for the Improvement of Flood Prevention (5-Punkte Hochwasserartikelgesetz) (Federal Act, 2004) is an example of progress made in legislation to promote integrated flood management and in particular flood protection measures. The essence of the Federal Act consists of the following five points:

- Joint flood protection program of the Federal Government and the Federal States.
- Cross-state action plans – International Professional Conference.
- Promote European Cooperation.
- Review river training works – develop inland navigation in an environment-friendly manner.
- Immediate actions for flood protection.

A particular point of the Federal Law is the adaptation of sector-laws to confirm with the objectives and directions of the Federal Act. Some of these laws adapted in the Federal Act include:

- Federal Building Code;
- Spatial Planning Law;
- Federal Water Way Law;
- Law of the German Weather Service;
- Law regulating Environmental Assessments.

Some of the policies adopted for Integrated Flood Risk Management can be seen in a wider policy framework to ensure the sustainability of flood protection measures.

In particular, sustainability policies can include (WWF, 2007):

- minimization of damage potential (reduction of value assets in high risk flood plains, removal of high value assets where feasible);
- ensuring improved spatial water retention especially in areas where floods originate from;
- development of new retention areas, especially through the backwards re-location of dykes;
- generating synergies between different policies and laws.

From past flood events, one of the most important lessons learnt is to further improve self-help capabilities of the potentially affected population, including flood risk awareness and best practices to adapt to and cope with risk. Included in self-help capabilities are flood-proofing of houses, taking flood insurances, and improving community driven evacuation procedures amongst other measures. Best practices can be adapted from experiences in other river basins in Germany, where the population is much more adapted to 'living with floods', such as in the Rhine and Mosel river basins. An important tool to sharpen flood risk awareness is the availability of a flood risk information system (http://www.zuers-public.de) where people can get information about the degree of flood hazards and risks. Such information is meant to sharpen awareness of environmental risks and serve as an incentive for self-help activities. In addition, German insurance companies use a geographical information system that allows the calculation of risks with regard to flooding, backwater storage and extreme precipitation.

The drive to strengthen self-help capabilities is critical in view of the increasing probability of the recurrence of flood events with a high damage potential and the scarcity of public funds and resources to pay for such damage in future.

Some of the successes in flood management during the floods in 2013 were not policy based but opportunity driven. The Task Team Reconstruction, Free State of Saxony (Germany), (Wiederaufbaustab, 2013) reported the outstanding role of social networks in the ad-hoc organization and implementation of flood management actions, and especially the rapid deployment of both government forces and volunteers to flood hot-spots where immediate help had been required. The important role of social networks the organization of flood management actions during flood events is fully recognized and plans are underway to make full use of them in support of information management as an important component of flood management during critical situations.

Interactions at the political level. As outlined above, responsibilities for flood risk management at the political level are distributed on federal, state and communal levels. This makes it difficult to agree on policies, especially as it is necessary to balance different interests and differences in the perception of which risks are acceptable and which are not. The positive side of this cumbersome process is that flood management in Germany is a highly interactive, participatory process involving many governmental and non-governmental agencies, academia, private industries and sector lobbies as well as the media and individual citizens with far-reaching rights to agree or not to agree to planned flood management measures in the planning phase. Often, it is the same people who are represented in different organizations and forums. All of this forms a very diverse community of stakeholders in flood management that allows for lateral thinking beyond sectorial interests and the finding of flood management solutions to which a majority can agree.

Communication and perception of risk. Clear communication of flood risk management is essential. At the local level, the importance of communication has been clearly recognized. Since May 2008, a web-based information service has been made available by the city of Dresden, showing interested citizens potentially flooded areas in high resolution. The maps show flooded areas without showing the effect of existing or planned flood protection structures. Therefore it is possible to also view areas that may be flooded if flood protection structures should fail (Dresden, 2011). On the communal level, 'flood-partnerships' have become more common in several federal states of Germany, where several communities along a stretch of a river align in a partnership to discuss risk awareness, levels of protection and actual measures.

Stakeholder involvement in flood risk management has the highest potential to minimize loss of lives and livelihoods and property. Practical measures include – amongst others – awareness and education programs including risk-based information such as using flood hazard and flood risk maps (Dresden, 2011), promotion of risk-avoidance and practical exercises including evacuation drills.

In the Elbe River Basin (and other river basins in Germany as well) the communication of residual risk is mainly achieved through risk awareness and risk-avoidance programs and training. Increasingly, flood hazard and flood risk maps have been tailored to visually demonstrate to the potentially affected population the areas and the degree of risk and also the residual risk, describing the consequences if (for example) a flood protection dyke fails as a result of an extreme event that is beyond the commonly used 100 years exceedance period.

Especially at community level, the 'As Low As Reasonably Practical' (ALARP) principle (Abhas *et al.*, 2012) provides a useful framework to identify different levels of risk acceptance including acceptable risk, tolerable risk and unacceptable risk from the viewpoint of societal risk perception. It is therefore a valuable component in stakeholder participation in decision-making processes at the lowest appropriate level of intervention. Examples are as follows:

- Acceptable risk: Flooding of extensively used agricultural areas; impaired access to other villages for a given time, impaired mobility of persons and goods.
- Tolerable risk: Flooded roads, minor flooding of houses without major damages to buildings and property.
- Unacceptable risk: Loss of lives, flooding of industrial sites with the danger of spilling chemical, toxic or radioactive substances; flooding of critical infrastructure including hospitals, power switching stations, etc.

However, this includes only 'first-order' risks. The risk perception of cascading risks as mentioned above is not well understood or reflected in flood management policies and practices.

The failure of flood protection infrastructure such as dykes is viewed as unacceptable but cannot be ruled out totally. This is described as residual risk. It needs to be noted, however, that German society in general is highly risk averse and what might be a tolerable risk in other countries may be viewed as unacceptable in Germany.

To ensure coherent risk management amongst riparian countries, appropriate means for communication are addressed in Chapter V, article 10 of the EU Flood Directive.

Inadequacies in policy development and implementation

As was mentioned above, clear communication of flood risk management is essential. In Germany, it is felt that policy makers need to provide clearer guidance as to which flood risk management measures are expected and which direction future protection should take (Zurich Insurance, 2014b).

There is an overall striving to achieve a truly integrated and applicable flood risk management policy in the Elbe River Basin. Much has been achieved, as outlined above. Due to a number of constraints – some of which are mentioned above – there are, however, inadequacies in policy development and implementation. Some of the root causes are outlined below.

One mainly political obstacle in the implementation of nonstructural measures in general, and the extension of flood retention areas in particular, is the widespread local resistance to the establishment of additional flood retention areas, including through the backwards re-location of dykes. This resistance is quite often based on the particular interests of those who would be restrained in their economic activities in areas that are declared as flood plains.

The Elbe Flood Protection Action Plan specifically mentions as a priority the backwards re-location of dykes and the re-activation of flood plains and, in part, this has been achieved. In summing up, however, the main thrust is still in investments in the reinforcement of dykes where some 45% of dykes are being reconstructed or reinforced, while the re-activation of floodplains amounts only to some 4–6% of those currently active (BUND, 2007). As part of the implementation of the IKSE Elbe Flood Protection Action Plan, dykes with a length of 513 km and reservoirs and polders for the improvement of flood retention in the order of 71 million m^3 have been constructed or reinforced (Socher, 2014). This shows that the majority of investment was allocated for technical measures in flood protection. In a wider context, the presently available flood plains amount to only 838 km^2 (14%) of the supposed original active flood plains of the Elbe River, with an estimated extension of over 6,100 km^2. Even if all currently planned backward re-locations of dykes were actually implemented, only 1% of the original flood plains would be re-gained (WWF, 2007).

Although the existing legislation is forward oriented with the aim to ensure effective integrated flood management practices, inadequacies exist in the applicability of specific laws and standing orders, as the complex legal system has numerous opportunities to avoid actual measures in flood protection and management. The main cause for this situation is that many laws have a wording that is not stringent enough and allows circumventing the actual intention of the laws to improve overall flood risk management. In this regard and in view of the ongoing discussions, positions of different stakeholders are documented. Some of them state that existing legislation is inadequately implemented in part because of an unabated industrialization (and intensification) of agriculture, infrastructure at risk in flood plains and exemptions to land use restrictions including building in high risk areas, etc. These cases are good examples for sector-oriented policies that are counter-productive for improved integrated flood risk management as described above. WWF (2007) stated that many existing laws and regulations are not formulated stringently enough to allow their enforcement. BUND (2007) concluded that inadequate policy integration (in German: *Politikintegration*) is a root cause for an overall deficient flood protection. In particular, the legal framework (such as the Federal Act for the Improvement of Flood Prevention passed in 2005 could not establish legally binding conditions important for flood protection as there are a number of exclusion rules or vague formulations that allow business as usual, such as building activities and intensive land use in denominated flood plains.

With regard to the sustainability of flood protection measures, a particularly critical issue is the heightening of dykes in response to design-flood reviews (which is a policy objective, as shown above) under a changing climate. While the immediate flood risk may be lowered and higher damages mitigated, the residual risk over a longer time horizon as a result of catastrophic extreme flood events increases in the case of a dyke failure. This would put even more lives, livelihoods and property at risk.

Public–private partnerships such as those between (local) governments, the private sector, the public and insurers could encourage risk reduction and risk financing (Zurich Insurance, 2014a). Such new alliances would need to be formulated in a policy framework. The actual forming of such alliances is still in its infancy.

Key issues in flood risk management

- The case of flood risk management in the Elbe has shown, from its inception, that a cross-sectorial formulation of an integrated flood risk management policy framework is indispensable to transform policies into action.

- These policies must be underpinned through an adequate legal system to make policies implementable and enforceable.
- The political will for cooperation at all levels and with a multitude of stakeholders, including public, private and academic institutions and organizations as well as riparian countries fosters the development of consensus-based sustainable policies.
- Resource requirements can be huge, which will require a long-term perspective of putting policies into action with multiple milestones established to monitor progress.
- Public flood risk awareness and largely improved flood risk information services including improved flood forecasting are key to keep flood damage to a minimum.
- Policy development is often based on projected future conditions and derived requirements for flood risk reduction. By themselves these policies are difficult to test for *a priori* sustainability and robustness. Likewise, most of these policies, once implemented, cannot be reversed easily because of the high costs involved. It is therefore advisable to verify policies in pilot projects such as those implemented in the ELLA (Elbe-Labe Flood Management Strategy) project (ELLA, 2005; Bundesanstalt für Gewässerkunde, 2006) and LABEL (Labe-Elbe Adaptation to Flood Risk) project (LABEL, 2012).
- Rebuilding efforts after floods without improved flood resilience are straining public resources. Closer linkages with the insurance industry and public–private partnership could improve this situation (Zurich, 2014a, b).

Research is another key issue that is often underestimated and, as a consequence of the Elbe flood in 2002, the national research program 'Risk Management of Extreme Flood Events' funded by the Federal Ministry of Education and Research (BMBF) was initiated (Burek & Rademacher, 2009). Between 2005 and 2010, 38 projects were carried out with a total value of 20 million euros (Peterson, 2009). However, there is a need for further research for flood risk reduction and an improved management of floods in a river basin context (CHR-KHR, 2004). It is astonishing that, to date, there is no clearly communicated 'research policy' addressing research requirements in support of policy development and implementation. Needs for further research include:

- systematic examination and verification of basic data-material including hydrological, meteorological and also socio-economic data;
- statistical analyses of trends and patterns in the hydrological cycle and flood events in particular;
- analysis of historical flood events;
- analysis and detection of changes or variability of weather patterns, cyclone tracks and related precipitation extremes;
- improved modeling of flood events from their genesis down to the propagation of flood waves along river stretches including tributaries;
- water-balance modeling of the catchment;
- scenario-based support to decision-making tools in support of policy development, adaptation and implementation including the effect of changes in areas such as: demographic change, land use (agriculture), infrastructure, climate change;
- analyzing of the retention effects of existing or projected flood retention measures (structural and non-structural) on particular flood events;
- emerging research needs may centre on improved information gathering and utilization of information such as that gathered through crowd sourcing.

Key policy messages for the regional and global community

- Policies need to be developed based on strategic targets and with a time horizon not less than 20 years.
- Policies should be developed building on the consensus of major stakeholders.
- Policy integration is of utmost importance: policies need to be developed and existing policies adapted at different levels across sectors in such a way that those policies are mutually complementary, non-contradictory and implementable to achieve improved integrated flood risk management.
- Flexibility of policies should be maintained so that they can be adapted to overall changing conditions.
- Policies should be anchored in legislative frameworks that can be enforced through by-laws, ordinances (standing orders) and supporting administrative procedures.
- Policies should be implemented through action plans and projects on local, provincial, national and international levels, as appropriate with the river basin as the planning and implementation unit. The long-term availability of critical resources (such as through the development of a joint (Federal and State level) financing strategy as mentioned above) needs to be ensured.
- Failure of policies in flood management can be very costly as a result of high investments and likewise high costs in the reversal of implemented actions on the basis of a failed policy. It is therefore advisable to 'experiment' with policies, such as through scenario-based impact studies and through dedicated pilot projects in sub-basins to ensure the sustainability and robustness of policies to be chosen.
- Development of policies to strengthen public private partnerships could go a long way to synergize efforts by the public and private sectors towards risk reduction approaches. Likewise, in this respect, a greater self-help capability (especially with a focus on a greater insurance penetration rate in the potentially affected population), if set as a policy, would aim towards alleviation of public expenses for flood damages while encouraging private partners to invest in building flood resilience through adequate measures (such as flood-proving houses).

In a FLOOD-ERA Report for England and Scotland on the 'systematization, evaluation and context conditions of structural and nonstructural measures for flood risk reduction' (CRUE, 2008), the authors came up with two main lessons. From their study on policy contexts their conclusions are cited *verbatim* as key policy measures, as these conclusions are fully applicable to the context of this paper with regard to the Elbe flood policies:

> *'The key lesson is that to fully comprehend flood risk management policy, and key policy decisions concerning flood measures, it is necessary to 'back up' into higher, broader level policies of Government where the roots of these policies exist and where the commitments to particular policy drives are formulated. High level Government policies may reflect grass-roots ideas and initiatives, but the strongest policy drive usually comes from the top downwards. Policies are translated into lower level policies via policy documents and signals emanating mainly from Government, with Government seeking to detect lower level concerns which may influence the policy agenda. A second key lesson is that in order to promote non-structural flood measures, it is often necessary in the first instance to inject germane ideas at much higher levels of policy.'*

Conclusions

The case of the Elbe River flood risk management efforts demonstrates that the development of coherent policies on the basis of stakeholder consultations is a promising way forward to transform policies into actions. Based especially on the relative success of the IKSE Elbe Flood Protection Action Plan, it can be shown that the development of flood risk management policies has to be seen as a process that includes international cooperation with riparian countries as well as cooperation between the German Federal States and the German Federal Government on one side and policy development at the level of the European Community on the other. The concept of policy integration in a cross-sectorial manner as well as embedding polices in a legal framework are key to make policies actionable. Non-structural measures have largely been left behind in policy implementation, in particular the 'room for the rivers' policy and backwards re-location of dykes in synergy with the environmental policy of ecologically healthy rivers and flood plains. Policy integration is of utmost importance: policies need to be developed and existing policies be adapted at different levels across sectors in such a way that those policies are mutually complementary, non-contradictory and implementable to achieve improved integrated flood risk management. Research policies are not visible in a way that research results would influence the development of policies and their implementation. Lessons learnt from the 2002 and 2013 Elbe floods also show that legal frameworks need to be more stringent at Federal and States level to improve policy implementation.

References

Abhas, K., Bloch, B. & Lamond, C. (2012). *Cities and Flooding – A Guide to Integrated Urban Flood Risk Management for the 21st Century*. The World Bank, Washington.

Belz, J. U. (2013). The German Elbe River Basin. Presentation at the MRC '*Extreme Flood Events – Past, Present And Future*' Conference, Bangkok (Thailand), 13–14 November 2013.

BUND (Bund für Umwelt und Naturschutz – Deutschland; Friends of the Earth, Germany) (2007). Hochwasserschutz an Donau, Elbe und Oder – aus den Fehlern nichts gelernt? Eine Bilanz des BUND im August 2007.

Bundesanstalt für Gewässerkunde (BfG) (1994). *The 1993/94 Flood in the Rhine Basin*. BfG Report No. 833 BfG, Koblenz.

Bundesanstalt für Gewässerkunde (BfG) (2006). Modellgestützter Nachweis der Auswirkungen von geplanten Rückhaltemaßnahmen in Sachsen und Sachsen-Anhalt auf Hochwasser der Elbe, Bericht BfG-1542, Koblenz.

Bundesanstalt für Gewässerkunde und Deutscher Wetterdienst (Hrsg.) (2013). Länderübergreifende Analyse des Juni-Hochwassers 2013. BfG-Bericht BfG-1797. Koblenz.

Bundesministerium des Inneren (2013). Bericht zur Flutkatastrophe 2013: Katastrophenhilfe, Entschädigung, Wiederaufbau, StabFH-51004/24#2.

Burek, P. & Rademacher, S. (2009). Veränderungen und Management der Risiken extremer Hochwasserereignisse in großen Flussgebieten – am Beispiel der Elbe (VERIS Elbe). Schlussbericht des Teilprojektes: Hydrodynamische Berechnung des Wellenablaufes der Elbe. BfG-1637. Bundesanstalt für Gewässerkunde.

Cash Online – Finanznachrichten, 23. July 2013; http://www.cash-online.de/versicherungen/2013/hochwasser-kostet-versicherern-fast-zwei-milliarden-euro/129.

CHR-KHR (International Commission for the Hydrology of the Rhine Basin) (1990). *Das Hochwasser 1988 im Rheingebiet*. Report No. I-9.

CHR-KHR (International Commission for the Hydrology of the Rhine Basin) (1999). *Eine Hochwasserperiode im Rheingebiet*. Extremereignisse zwischen Dezember 1993 und Februar 1995, Report No. I-17.

CHR-KHR (International Commission for the Hydrology of the Rhine Basin) (2004). *Development of Methodologies for the Analysis of the Efficiency of Flood Reduction Measures in the Rhine Basin on the Basis of Reference Floods*. Report No. I-21.

CRUE (2008). *Systematization, Evaluation and Context Conditions of Structural and Non-structural Measures for Flood Risk Reduction; FLOOD-ERA Report for England and Scotland.* CRUE Research Report No. I-1.

Deutscher Bundestag (2013). Bericht zur Risikoanalyse im Bevölkerungsschutz 2013; Deutscher Bundestag Drucksache 18/208 18. Wahlperiode 16.12.2013.

Dresden (2011). Potenzielle Überschwemmungsflächen der Elbe jetzt online im Themenstadtplan verfügbar. Press Release. City of Dresden (5 May 2011). See: http://www.dresden.de/de/02/035/01/2011/05/pm_020.php (accessed 29 August 2014).

EC (European Commission) (2000). Directive 2000/60/EC of the European Parliament and of the Council establishing a framework for the Community action in the field of water policy (EU Water Framework Directive).

EC (European Commission) (2004). Communication from the commission to the council, The European Parliament, The European Economic and social committee and the committee of the regions flood risk management flood prevention, protection and mitigation.

EC (European Commission) (2007). Directive 2007/60/EC of the European Parliament and of the Council of 23 October 2007 on the assessment and management of flood risks.

ELLA (2005). *ELBE – LABEL Preventive Flood Management measures by transnational spatial planning (ELLA).* Saxon State Ministry of the Interior; http://www.bafg.de/DE/08_Ref/M2/03_Fliessgewmod/elbe_rueckhaltem_hw.pdf?__blob=publicationFile.

European Flood Action Programme (2014): http://ec.europa.eu/environment/water/flood_risk/com.htm.

Federal Act (2004). Gesetz zur Verbesserung des vorbeugenden Hochwasserschutzes Deutscher Bundestag – 15. Wahlperiode Drucksache 15/3168.

Heintz, M. D., Hagemeier-Klose, M. & Wagner, K. (2012). Towards a risk governance culture in flood policy—Findings from the implementation of the 'Floods Directive' in Germany. *Water*, *4*, p135–156. doi: 10.3390/w4010135

IKSE (Internationale Kommission zum Schutz der Elbe; International Commission for the Protection of the Elbe River) (2003). Aktionsplan Hochwasserschutz – Elbe (The Elbe Flood Protection Action Plan). Internationale Kommission zum Schutz der Elbe, 39006 Magdeburg Germany.

IKSE (Internationale Kommission zum Schutz der Elbe; International Commission for the Protection of the Elbe River) (2012a). Abschlussbericht über die Erfüllung des Hochwasserschutzplans Elbe 2003–2011 (International Commission for the Protection of the Elbe River). IKSE, 39006 Magdeburg, Germany. http://www.ikse-mkol.org/index.php?id=758&L=0%2220onfocus%3D%22blurLink.

IKSE (Internationale Kommission zum Schutz der Elbe; International Commission for the Protection of the Elbe River) (2012b). Länderübergreifender Hochwasserschutz an der Elbe - Ein Abschlussbericht 10 Jahre nach der Flut. IKSE, 39006 Magdeburg, Germany. http://www.ikse-mkol.org/index.php?id=1&L=2.

IKSE (Internationale Kommission zum Schutz der Elbe; International Commission for the Protection of the Elbe River) (2013). Das Hochwasser im Juni 2013 und der internationale Hochwasserrisikomanagementplan im Einzugsgebiet der Elbe, workshop (21 November 2013). IKSE, 39006 Magdeburg, Germany. http://www.ikse-mkol.org/index.php?id=852&L=.

IKSR (Internationale Kommission zum Schutz des Rheins; International Commission for the Protection of the Rhine) (1998). *IKSR – Aktionsplan Hochwasser.* (Rhine Flood Action Plan) Report No. 156. See: http://www.iksr.org/index.php?id=123.

ILRI (International Livestock Research Institute) (1995). *Livestock Policy Analysis*. ILRI Training Manual 2. ILRI, Nairobi, Kenya. pp. 264.

Koalitionsvertrag (2014). Deutschlands Zukunft gestalten: Koalitionsvertrag zwischen CDU, CSU und SPD, 18. Legislaturperiode, Kapitel 4.2, „Umwelt' p. 120; http://www.bundesregierung.de/Content/DE/_Anlagen/2013/2013-12-17-koalitionsvertrag.pdf;jsessionid=BE45E5D27F84A05F7B6594747EFE0CD6.s1t1?__blob=publicationFile&v=2.

LABEL (2012). *LABE-ELBE Adaption for Flood Risk (LABEL).* Saxon State Ministry of the Interior. http://www.label-eu.eu/uploads/media/EN_LABE-ELBE_2012plus.pdf.

LAWA (2014a). German Working Group on Water Issues of the Federal States and the Federal Government. http://www.lawa.de/About-LAWA.html.

LAWA (2014b). *Zusammenfassende Analyse der Ergebnisse der vom Hochwasser 2013 betroffenen Flussgebietsgemeinschaften.* Bund/Länderarbeitsgemeinschaft Wasser (LAWA).

Peterson, D. (2009). Sustainable sanitation and water management, Creating Policies and a Legal Framework. Seecon international GmbH; http://www.sswm.info/print/1605?tid=596.

Plate, E. (2011). In: Large-Scale Floods Report, ICHARM Book Series No. 1.

Pressespiegel (2014). Elbe-Anrainer fordern mehr Raum für Gewässer. http://www.mdr.de/nachrichten/elbe-anrainer100.html.

Raadgever, G. T. (2005). Transboundary River Basin Management Regimes: The Elbe Basin Case Study, Background Report to Deliverable 1.3.1. of the NeWater project, RBA Centre, Delft University of Technology, Delft.

Rickmeyer, A. (2014). Konsequenzen aus den Hochwasserereignissen 2013 Niedersächsisches Ministerium für Umwelt, Energie und Klimaschutz 23. April 2014 - BWK Hannover.

River Basin Community – Elbe (http://www.fgg-elbe.de/start-en.html).

Rudolph, K. U. & Block, T. (2001). *The German Water Sector – Policies and Experiences*. Federal Environmental Agency, B., Germany.

Socher, M. (2014). Informationsveranstaltung 'Hochwasserschutz an der Elbe'; 1 | 5. Juni 2014 Bad Schandau, Presentation.

Vanneuville, W., Deckers, P., Van Eerdenbrugh, K. & Mostaert, F. (N.N.) Flood Risk Scenario Calculations as a Decision Support and Evaluation in Water Management Plans. www.vliz.be/imisdocs/publications/135615.pdf (accessed 9 September 2014).

VERIS – Elbe (2013). Changes and Management of Risks of Extreme Flood Events in Large River Basins – The Case of the Elbe River. http://www.veris-elbe.ioer.de/template-en.php?bereich=projekt (accessed 30 August 2014).

Wiederaufbaustab (Reconstruction Team), State of Saxony (2013). Der Wiederaufbau im Freistaat Sachsen nach dem Hochwasser im Juni 2013.

WWF (World Wildlife Fund) (2007). 5 Jahre nach der Elbe-Flut; Editor: WWF Germany, Frankfurt/Main.

Zurich Insurance (2014a). European Floods: Using Lessons Learned to Reduce Risks; www.zurich.com/_/.../european-floods-using-lessions-learned.pdf?la=en.

Zurich Insurance (2014b). Risk Nexus – Central European Floods 2013: A Retrospective.

Lessons learned from the Big Floods, 1999–2013: The Bavarian Flood Action Programme 2020plus

Martin Grambow*, Erich Eichenseer, Gregor Overhoff, Tobias Hafner, Kerstin Staton and Bernhard Schaipp

Bavarian State Ministry of the Environment and Consumer Protection, Rosenkavalierplatz 2, München 81925, Germany
**Corresponding author. E-mail: martin.grambow@stmuv.bayern.de*

Abstract

Flood protection is an integral part of society's development. Meeting a growing vulnerability in the context of geopolitical and global changes, existing systems of natural hazard management must be reviewed on a regular basis. This is illustrated well by Bavaria's recent history. The lessons from the big floods in the period of 1999–2013 led to a fundamental modification of Bavaria's integral flood protection strategy. The concepts of flood risk management, resilience and dealing with extreme flood events which exceed standard structural design limits came to the fore. Management of flash floods and other floods, handling of potential retention areas, the inclusion of insurance, resettlement, and the burden of maintenance are further challenges which are briefly addressed in this paper.

Keywords: Bavaria; Danube; EU Floods Directive; Flood protection; Maintenance; Resilience; Risk management; Torrents

1. Introduction

1.1. Water as a public good and a basic public safety issue

Water resource management touches upon basic safety issues: drinking water and nutrition, water as a hygiene factor, water as an eco-system and especially as a natural hazard posing a direct threat to humans (as well as to all the aforementioned water issues).

As a gravimetric element, it links all of society with an array of circumstances upriver and downriver alike. Water comes to everyone. Everyone can benefit from good management and suffer from poor management. That is why water management in general and especially flood protection are special

doi: 10.2166/wp.2015.007

essential public services, and a component of the basic protection strategy falling under a state's fundamental responsibility. That is why, above all else, the duty of those involved in politics and administration is to derive strategies to the best of their knowledge and abilities, that will be fair and preventative in nature, in the spirit of sustainable policies.

Doing so, however, involves both centralised and decentralised components. A state's water management is subject to both national and international frameworks (in this case, German and European), as well as regional and local ones. Within the scope of good governance, execution of the strategy will always have components that are both of a sovereign/public nature, i.e. resting on the state and community, as well as of a private sector nature, i.e. private industry, institutional and individual. The reality, of course, is always on site.

1.2. Outstanding characteristics of EU water policy

Attempts to provide flood protection are as old as hydraulic engineering itself. Bavaria has been investing in natural hazard management for more than 200 years within the scope of strategic state water management. There have always been floods, yet the striking frequency of the events from 1999 to 2013 after nearly 30 years of having only a few flood events has revealed weak points. Not only has the number of protective structures been increasing due to ongoing investments, but also our technological society has apparently grown more sensitive, with damage constantly rising year on year as put forward by insurance organisations and statistics (MunichRe, 2014). The time series on which flood protection structural design is based are comparatively brief, with uncertainties additionally being cast due to climate change predictions.

What is helpful is the great amount of knowledge gained as a consequence of the establishment of European norms. To that end, the EU Water Framework Directive (2000) and EU Floods Directive (2007) have yielded integral approaches representing an outstanding basis for sustainable water management. However, that does not mean that we have learned all we need to know – on the contrary!

2. Brief description of Bavaria

Bavaria is a water region, and more than 100,000 km of streams and rivers flow through it (Table 1). Bavaria is internationally active in water management.

The State of Bavaria actively fosters mutual water policies in Europe as a member of international water commissions (Danube, Elbe, Rhine, Lake Constance) and consortia (Arge Alp, Arge Alpen-Adria). Moreover, Bavaria's water economy recognises its obligation to pass on its knowledge and experience not only within the context of these international organisations, but also in bilateral projects with states and regions interested in the model shaped in Bavaria of integral state water management based upon sustainability criteria. This obligation is simultaneously an opportunity: an opportunity to implement practice-proven Bavarian environmental technology in order to improve environmental and living conditions in these regions, thereby garnering mutual experience while also securing and enhancing Bavaria as a home for business and science.

Table 1. Key numbers concerning the Bavarian water region.

- Area of Bavaria 70,550 km^2
- Gross Domestic Product (GDP) (as of 2013): €488 billion[1]
- Residents (as of 31 December 2011): 12,595,891
- Administrative communities: 2,056
- total length of all bodies of water (rivers): approx. 100,000 km, of which first-order (large, trans-regionally important) 4,200 km; second order (regionally important) 4,800 km
- Number of larger natural lakes: 150; total area: 270 km^2
- State dams and flood control basins: 25
- Hydro power plants: approx. 4,200
- Water cycle: average annual precipitation in Bavaria: 939 mm (south max: 2000 mm, north min. 450 mm), runoff 422 mm, evaporation 517 mm
- Highest point: Zugspitze at 2962 mNN (metres above sea level); Lowest point: Main 102 mNN, Danube 280 mNN
- Public water supply (as of 2010): percentage connected 99.1%, potable water protection areas approx. 3,250, i.e. ~4.6% of Bavaria's area (national territory in Germany: ~12%)
- Waste water disposal: 96.4% of residents have a connection to the 2,669 waste water treatment plants (as of 2010)
- Hundreds of hydrography measurement points along surface waters and in the ground water for monitoring level qualitatively and quantitatively

3. Flood cluster 1999–2013

The period from the 1960s to the 1990s is considered in Europe as well as in Bavaria to be a period with few flood events. There were, of course, extreme local events of HQ100[2] dimensions, but no sweeping catastrophic floods. That changed in Bavaria with the flood of 1999, where heavy rain in the Alpine foothills caused heavy damage, e.g. along the Danube after the dyke breach near Neustadt. In 2005, billions of euros of damage were caused again due to rain, this time in the western Alpine region. In early June 2013, vast areas of Bavaria, this time both the Alpine region and northern Bavaria were hit by a massive flood event causing around €1.3 billion of damage (Figure 1). The last flood was an extreme event with regard to its duration, geographic coverage and exceedingly high peak discharges. At some locations, the peak discharges were the highest ever recorded and reached or even surpassed the structural design limits of the flood protection systems.

As is often the case, this large-scale event occurred in the 'third wave':

- In May 2013, 25 days of rain with an aggregate high level of relatively constant rainfall led to a high level of pre-saturation in the soil. Local flooding was then followed by flood water on a moderate scale.
- Then came further intensive rainfall throughout all of eastern Bavaria from 30 May to 3 June, especially in the Alps, at times corresponding to a rain event well over a 100-year event for durations of 72–96 h.

The water level rose sharply, especially in the southern Danube basin, at times above the level of a 100-year event. Thus, only a relatively short period of time after the floods of 1999, 2002 and 2005,

[1] Billion = $\times 10^9$.

[2] HQ100 describes a flood with a recurrence probability of 100 years: 'a 100-year flood'. This is the standard design flood in Bavaria.

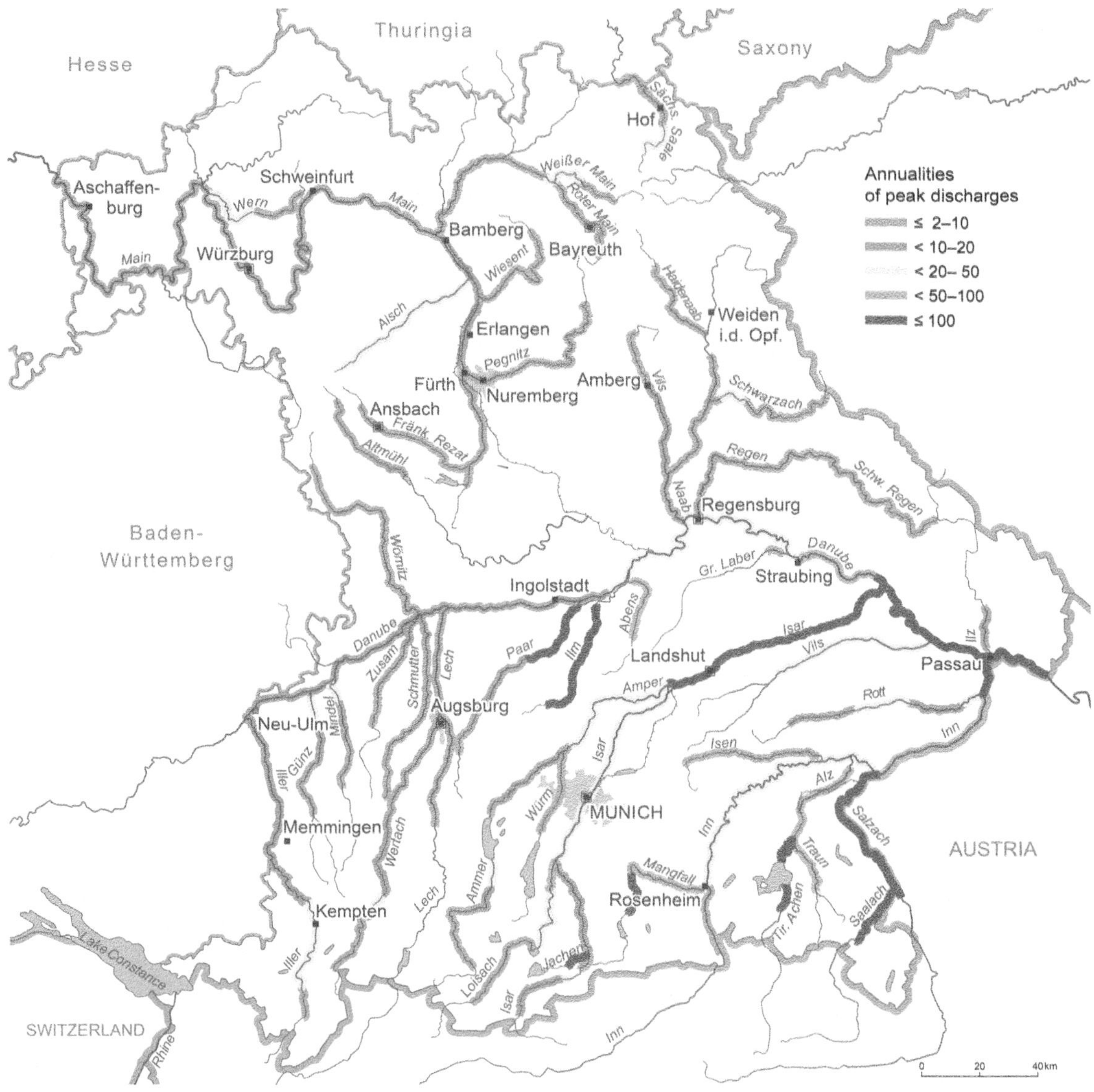

Fig. 1. Affected river sections in the May/June 2013 flood.

another catastrophic event occurred. While the geographical focal points of the events do vary greatly, this series of events can indeed by classified as a flood cluster which may even be marked by weather changes caused by climate change.

4. Lessons learned following the large-scale floods of 1999–2013

Despite all the damage and problems, the existing flood protection installations in place in Bavaria fundamentally proved themselves. The investments made over past years were well placed. This made it possible to prevent damage of a much larger scale; the Bavarian administration estimates the ratio of investment in flood protection to damage prevention to be ~1:4 to 1:7. Even when taking capital costs into consideration, the financial calculation remains positive, not to mention the prevented human suffering – floods traumatise people in a significant manner.

An assessment of flooding consequences reveals a few significant characteristics:

- In 2005 and 2013, many systems – especially dykes – were subjected to significant overload (flood discharge >> structural design discharge). Many of these systems successfully withstood this overload due to high safety design margins. Other systems, although overrun, still withstood the overload and significantly helped the mitigation of the damage. Several systems, however, did fail and were completely eroded.
- All flood protection systems meeting today's flood design standards and providing protection against a 100-year event withstood the 2013 events. However, this was frequently the case with full exhaustion of safety heights and occasionally structural damage to the system.
- Damage was largely incurred along river sections that were not yet protected or not sufficiently protected: either because installations were not yet developed to HQ100 or because the river was calculated to withstand the HQ100 discharge (and was then overloaded by an HQ 300–500). The worst damage was the result of sudden and/or complete (eroding) failure of flood protection installations.
- In 2005 and 2013, the controlled retention areas (especially state reservoirs) made a large contribution to damage prevention. At approximately 100 million m^3, the Sylvenstein Dam alone helped prevent billions of euros of damage in Munich.
- The collaboration between all parties (administrations and crisis intervention staff) throughout the course of the event was excellent. Good weather forecasts and solid preparation contributed to that end.
- There were minimal cases of deficits in flood predictions and flood warnings, where real conditions 'overran' model scenarios, or where catastrophic events destroyed river gauges or cut-off power supplies, thereby cutting off communication.

The extreme flooding seen in June 2013 once again explicitly underscored the importance of having modern and effective flood protection in order to protect the population and avoid economic damage, additionally illuminating how decisive a society's systemic ability to react is. In the aftermath of the flooding, the tremendous solidarity and willingness to help, seen throughout Bavaria and Germany was remarkable. €8 billion in assistance came for the federal and state government for rebuilding.

Statements that can be made as a first conclusion based on recent experiences include the following:

- Statistically calculated flood events (design discharge) occur irregularly, yet they certainly do occur. Timely investment and preparation avoids enormous consequential costs.
- Mega events that far exceed HQ 100 can occur. Despite all endeavours carried out with technical protection systems and preventative measures, safety against flooding is always only possible to a limited extent. A residual risk will always remain and cannot be remediated regardless of the expenditure made.
- This residual risk can only be reduced through individual action: assuming individual responsibility in making provisions and correct action in a crisis situation can make a significant contribution to

reducing damage. Modern integrated flood protection and flood risk management is therefore a duty spanning across multiple departments and fields (development planning, settlement policy, location and risk mitigation in business and agriculture, the insurance industry).
- After events of this nature, it is necessary to fully evaluate and improve the flood prediction models implemented. Interfaces are found in precipitation forecast based on location and volume, discharge measurement during extreme events and cross-border collaboration.
- The interplay among the four principle factors contributing to river discharge (precipitation, runoff, channel hydraulics and wave superposition) is partially non-linear and cannot be determined with intuitive cause and effect presumptions. This is especially recognised in the philosophical dispute between 'natural' vs 'technical' flood protection.
- Instead of pure protection, it is important to think about the *resilience* of protection systems in the face of extreme events. The focus here is on systemic and precise overload stability (system and subsystems).
- Flooding has illustrated once again that flood protection systems not only have to be well planned and executed, but also well maintained. Ensuring they are in a solid, functional state is a permanent task of growing dimensions.
- Quality in flood protection is yielded by the sum of the qualities found among the personnel involved.

Based on this information, several fundamental positions and insights will be examined in detail for the purpose of illustration.

4.1. Universal strategy in dealing with flooding

For more than 200 years, state flood protection has been actively carried out along Bavarian rivers in line with modern criteria (Wiebeking, 1817). New information and technical advancements have triggered more and more adjustments in the fundamental protection strategy. After the flood during Pentecost 1999, the Bavarian Cabinet enacted the 'Flood Water Action Programme 2020 – for sustainable flood protection in Bavaria'. It lays out a specialised strategic orientation for flood protection, as well as a framework for finances and time for implementation. The Action Programme 2020 (AP 2020; see: http://www.lfu.bayern.de/wasser/hw_aktionsprogramm_2020/index.htm) follows an integral, pioneering protection strategy consisting of three (equal) fields of action:

- natural retention;
- technical flood protection; and
- prevention and extended precautionary action against flooding.

This means that it represents a new level of quality in integral protection strategy all over Bavaria. It spells out investments of €115 million per year, to a total €2.3 billion. This has made it possible to protect an additional population exceeding 400,000 residents along first-order and second-order bodies of water against 100-year floods. Many of these measures have already proven their value multiple times and avoided enormous damage on multiple occasions. Several of the implemented measures displayed their effectiveness in an impressive manner during the flooding of June 2013.

This three-pillar strategy continues to be current and also facilitates the implementation of recent knowledge gains. The enhancements described below largely pertain to enhanced system observations, the inclusion of complex, non-linear factors and better quantification of influential factors that are frequently only made possible by modern computing.

4.2. Flood risk management

For many generations, people spoke of 'flood clearing'. This term is no longer suitable in light of current knowledge regarding the systemically established limits to active protection against natural hazards. Flood risk management is aimed at organised/systematic and balanced handling of flooding that significantly reduces damage with tailored utilisation and action in consideration of the residual risks. Specifically, effective handling of flood water means more than simply setting up technical protective structures. Rather, it is necessary to develop an integral strategy and also to take into consideration aspects like effectiveness, efficiency and suitability against the background of probability (and unpredictability) of rare extreme events and the resulting overload on protection systems. Another aspect to flood risk management of this kind is that various fields collaborate in a cross-border and multi-disciplinary manner. Examples of the fields involved here include water management, land use, nature conservation, local hazard control and disaster prevention.

Flood risk management encompasses all phases before, during and after flooding. It can therefore be understood as an ongoing cycle dependent upon time (Figure 2).

Thus, addressing flood risk is not a one-time task, but rather a process consisting of various phases that have to be run through on a regular basis.

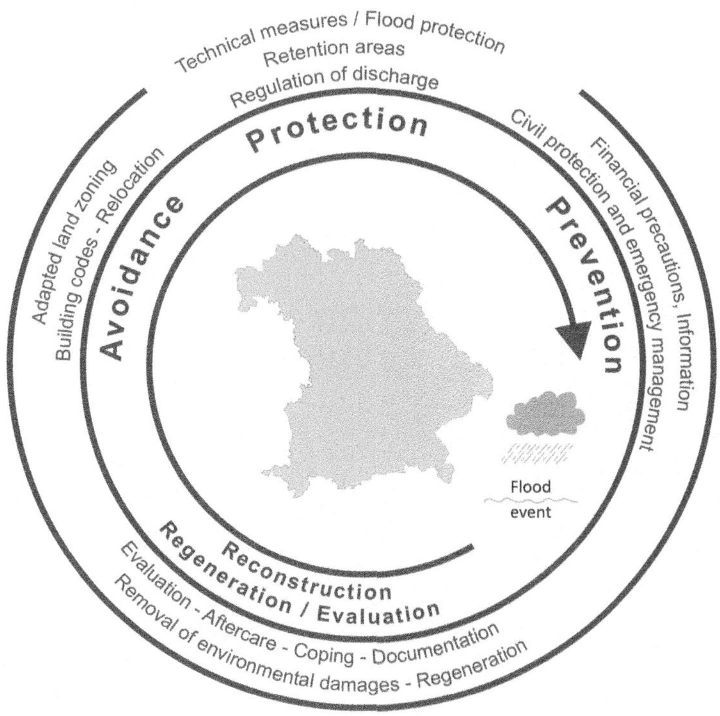

Fig. 2. Flood risk management cycle.

The revision to the first flood risk management plans for Bavaria stipulated in the EU Floods Directive (2007) was designed to introduce and permanently establish systematic flood risk dialogue. Only those who are aware of the hazards and risks will be capable of implementing effective provisions and acting correctly in the event of a serious incident. This applies for all actors – from state, city and community governments, to those responsible for infrastructures covering several regions, to industrial and commercial operations, and each individual citizen.

The foundation for constructive risk dialogue consists of maps visualising the hazards and risks to the public in a comprehensible manner. Risk dialogue is carried out on a local level between cities and communities, water management agencies and district administration offices in their capacity as disaster prevention agencies.

4.3. Tailoring the strategy to Bavaria – from AP 2020 to AP 2020plus

In light of the flood events in June 2013, Bavaria re-adjusted the cornerstones of flood water protection in Bavaria with the current 'Flood Water Protection Action Programme 2020plus' (AP 2020plus). AP 2020plus is a logical continuation of AP 2020 based on flood risk management. This is expressed in Figure 3, which integrates AP 2020plus with the risk management cycle.

4.3.1. Organisational adjustments. To start, the terms 'risk' and 'risk cycle' better reflect legislative developments from the EU Floods Directive (2007), which was implemented into national law in 2010.

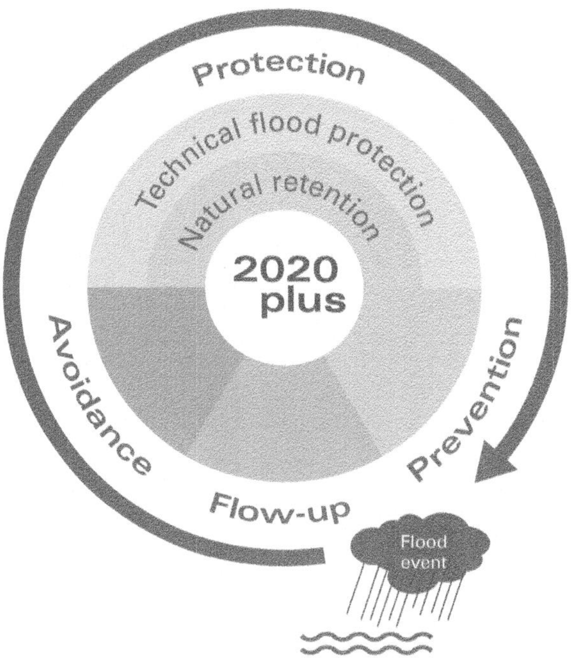

Fig. 3. Elements of the Action Programme 2020plus.

Another element of AP 2020plus is better financing and staffing. In the future, there are plans to provide €150 million instead of €115 million for implementation of AP 2020plus. Additionally, plans were passed in February 2014 for a special programme for flood protection on the Danube featuring an aggregate €600 million. When accounting for the higher expenditures made to date (compared to the €115 million/year planned), this means that the AP 2020 and AP 2020plus financial volume for flood protection is increasing from €2.3 billion to €3.4 billion. Moreover, reductions in flood protection staffing have been halted, with an initial 150 new positions created – to be staffed up to 2022.

One of the big challenges for the future consists of the large inventory of flood protection installations and dams, as well as structures designed to protect against torrent hazards. The replacement value for state hydraulic engineering structures in Bavaria is estimated at €10 billion. These installations require regular monitoring, maintenance and, when necessary, restoration. This overall required effort grows with each new structure.

4.3.2. Technical/operational improvements. Additionally, AP 2020plus lays out technical/operational improvements and enhancements in all fields of flood risk management. Select examples are:

- further improvements in notification and warning services (www.naturgefahren.bayern.de) including improvement in flood forecasting;
- investing in risk dialogue (in line with implementation of the EU Floods Directive);
- adoption of design and construction of protection systems also based on resilience criteria;
- determination of flooding regions in further river sections.

The following will address and detail the core points in the technical/operational adjustments to protective measures.

4.3.3. Overload and systemic reinforcement of resilience. The June flooding displayed once again the catastrophic consequences of dyke breaches, especially along the Danube. The affected dykes there were not designed to accommodate a 100-year event and were overloaded. However, this particular example also makes it clear that flood protection has its limits and that there will always be a residual risk. Overload can occur anywhere any time, regardless of the statistical flood parameters used to design and build protective structures; an even larger event can occur at any time. In addition to 'the ability to resist external influences', the term resilience also has more complex characteristics. Over time, systems typically develop through growth and are stable up to a certain point. After this point a sudden decline occurs, if the decline is not prevented via early intervention and renewal – an allegory for the upkeep of installations. When systems transgress their threshold, they fall into a new stable state: a dyke is flooded, eroded and 'disappears'. Ultimately, all systems are comprised of superordinate and subordinate sub-systems. The resilience in each sub-system corresponds to the superordinate and subordinate sub-systems, i.e. it is influenced by them while also influencing them (Walker & Salt, 2006).

The lesson to be learned here for flood protection is that, in the event of overloading, there will be further influences, such as ageing of systems, sub-optimal control and criminal acts rendering it impossible to completely prevent damage. However, it makes a huge difference whether an installation totally fails and collapses in the event of overload, or survives the overloading and thus continues to provide a partial protection function. Taking a dyke as an example: the damage is much lower when the water level behind the dyke (polder) rises only slowly with lower momentum, giving additional warning time. In addition, after flood wave peak(s), overflow into polders ceases quickly when dykes are still in place. Ultimately, the degree of protection is maintained during subsequent flood waves, and the polder is not left defenceless.

The resilience of one single sub-system or of the superordinate system (e.g. river section, body of water) can be improved through constructive measures on various planning levels, with construction details to the aggregate system (Figure 4). The following mutually complementing options come into consideration.

- Structural safety: selection of fundamentally overloadable and therefore resilient construction methods and structures, e.g. flood protection walls, anti-erosion dykes (e.g. sheet piling and geotextiles) and dyke sections designed to allow overflow. When it comes to dams, these kinds of relief systems

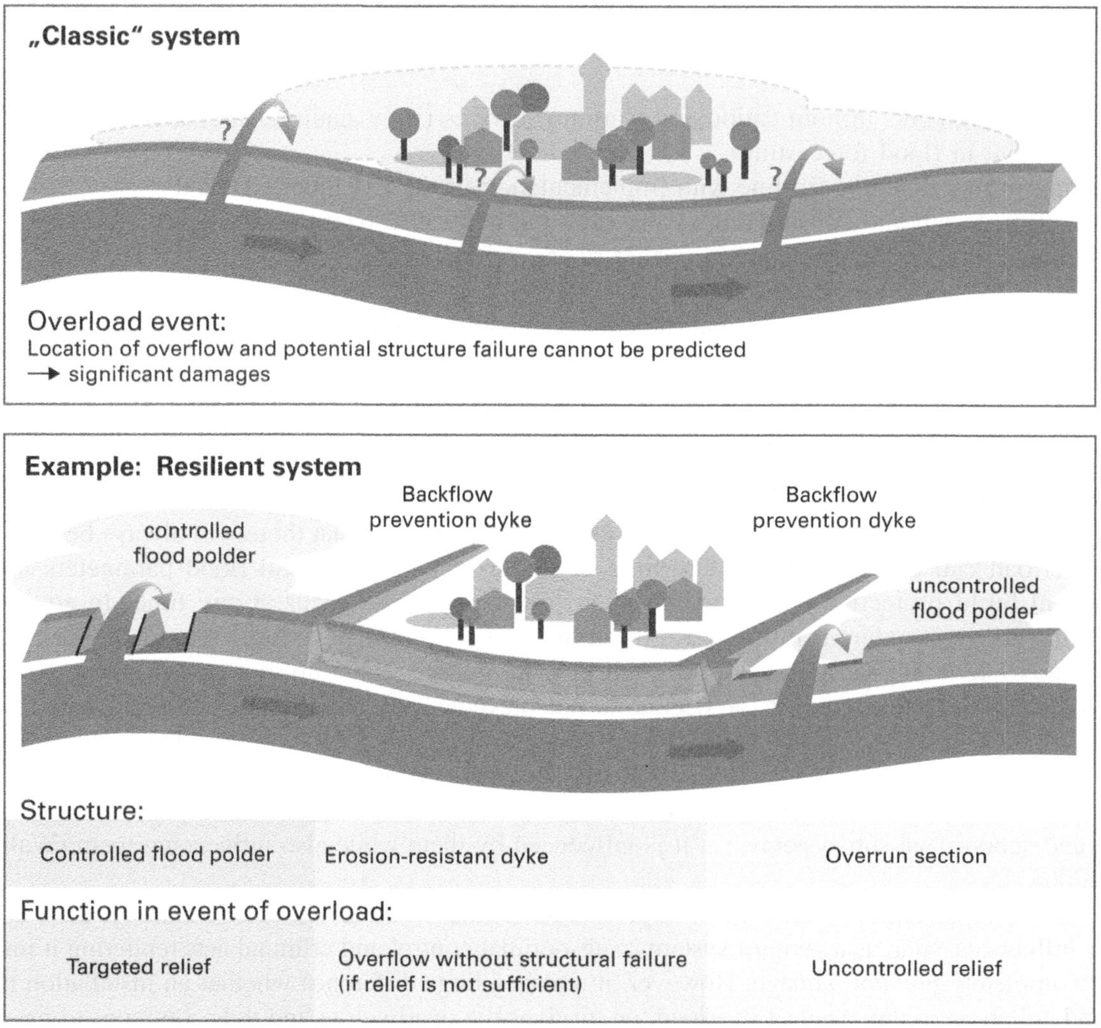

Fig. 4. Principal outline for potential components of resilient protection systems.

(spillways) have long been standard due to their better ability to take on damage. However, this also includes redundant power supply, capacity for manual operation and sufficient personnel with regards to numbers and training.

- Systemic safety: conceptualisation of overloadable and therefore resilient protection systems; the resilience of the individual structures is complemented with supplemental elements, such as natural retention spaces, relief sections, flood polders and further dykes (transversal dykes). This makes it possible to either retain a design flood longer, or at least to provide partial relief due to volume reduction.
- In combination with the 15% climate increment on the HQ100 design discharge introduced throughout all of Bavaria back in 2005, safety has significantly increased in situations involving design flood levels and beyond.
- The residual risk remaining can be further reduced with management measures. It can be reduced in these areas with further prevention measures, e.g. through personnel planning in disaster prevention organisations, voluntary adjustment to utilisations and elemental damage insurance. Warning systems include stable forecast models, anti-flood measuring and monitoring systems as well as the preparation of individual decision channels in the event that central management is lost (contingency strategy for management, etc.).

It is under these conditions, and only under these conditions, that the elevated degree of utilisation necessarily yielded by the improved protection level will not result in greater systemic susceptibility.

4.3.4. Installation safety and resilience. Flood protection installations are generally complex systems built using cement, steel and/or earthworks requiring substantial expenditure for planning, construction and operation. Before these structures are built, extensive planning and legal procedures are usually carried out and financial issues are clarified, generally involving contributions from the communities and cities being protected.

Upon completion of the installations, the citizens in question expect an extremely high level of protection, in most cases absolutely fail-safe protection against flood hazards. Since safety installations only seldom have to display their suitability for use, and when they do it is during flood events, it is not possible to perform a long, staged 'trial back-up' to observe the structure's performance and discover potential weak points requiring improvement (one exception to this being dams and flood detention reservoirs subjected to systematic load increases under 'normal' discharge conditions via a trial back-up programme).

Thus, flood protection installations need to be constructed such that they deliver a high degree of inherent safety. This can be achieved through the selection of building materials, through reserves in load assumptions, through sufficient freeboard measurement, and through subsidiaries for hydrological parameters (e.g. in Bavaria, 15% increment on HQ100), etc. This makes it possible to render earthwork structures, at risk of being overrun in the event of overload, resilient with surface erosion protection or with additional sealing within the dyke (earth/soil concrete, diaphragm wall, sheet pile wall), thereby providing protection against sudden structural failure. The relief options laid out in the relevant norms regarding protection of the aggregate system against sudden failure (intake structures) are generally 'secured by sealing them with sand bags' (not intended when designed and thereby robbed of their function). This makes resilience all the more important, i.e. the ability of flood protection installations to provide protection against failures in the sense of the threat of overload. Damage to installations and protective materials is unavoidable even when subject to loads accounted for in the design. What is important here is that sudden failure in protection systems and the corresponding catastrophic consequences can be hindered and that the flood levels remain under control.

In contrast to classic flood protection installations, dams and large river barrages represent a constant potential hazard with their high permanent water level above the surrounding terrain. That is why these large-scale water installations are built with particularly reliable design, equipped with safety margins laid out in the current applicable norms (design discharge, freeboard, failure in outlet culverts (n-1 case)) or are retrofits of older installations as needed, operated by specialist personnel and carefully monitored. Timely assessment of measured values and observations, separation between in-house and third-party monitoring, annual safety reports on the current state of the installation including any changes, as well as detailed checks of the entire installation approximately every 15 years guarantee a high level of safety in Bavaria's dams and water reservoirs.

When abnormalities in installations are recognised, punctual counter measures are implemented in order to maintain the overall high safety level. The State of Bavaria makes considerable expenditures to that end. Dedicated retrofit projects are contracted for older rivers dams and have been implemented for the first time in order to bring those dams back into the desired state without interrupting operation or restricting functionality. One example for these kinds of measures is the successful construction measures on the Sylvenstein Dam, carried out from 1994 to date (the retrofit of a pore water pressure monitoring network in the dam's body, construction of a second spillway, dam raising by 3 m, construction of a diaphragm wall in the dam core, insertion of an inspection gallery, and construction of a new seepage measurement system inside the dam).

Big gains are consistently being made in knowledge regarding technology and systems in all fields. Paradoxically, this knowledge is also shedding light on the recognised requirements. An evolution in efficiency alone is no longer sufficient, especially when doing so removes redundancies that were indispensable to the system's resilience. This is why research and development needs to be supplemented with components for sustainable system stability.

4.3.5. Effects on discharge in large rivers. The underlying concept of Integrated Water Resources Management is to observe the whole drainage basin as an effects matrix. Through observation and calculation, we have now learned that the runoff taking place within a typical drainage basin is anything but trivial:

- Rainfall is difficult to forecast and varies in intensity, duration and pull direction.
- Runoff is influenced by the surface structure as a function of gradient, relief, vegetation and infiltration rate or surface sealing – here too, times are dependent upon vegetation, frost and rate of saturation from past rain.
- In the main channel flow and the forelands the discharge is influenced by gradient, discharge coefficients and retention capacities.
- One component that is typically non-linear consists of the estuaries of larger tributaries.

The sum of the non-linear overlapping in these four transient and dynamic components makes predicting effects extraordinarily difficult. Even generally evident assumptions such as 'more retention area is always good' and 'discharge acceleration is always bad' may be incorrect. Taking the Danube

River near Passau as an example, discharge delays due to retention can indeed lead to undesired wave overlapping with the Danube's normal wave.

Thus, it can be presumed that every river system has its own character, even when properties such as the drainage basin and runoff appear to be similar. In order to make statements regarding the effect of flood protection measures beyond the general hydro-engineering assumption, the Bavarian Danube (Bavarian watershed area of 4,047 to 77,023 km^2, HQ100 from 513 to 8800 m^3/s) was modelled, including its historical, current and potential retention space using complex two-dimensional hydraulics. This is possible today because both geo data as well as hydraulic modelling are resolvable with state of the art technology. One preliminary result of these studies – which are still underway – on a 260-km stretch of the Danube river between the Iller estuary and Straubing (Rutschmann, 2013) is that several established assumptions have been confirmed and that new knowledge has been generated:

- Reactivation of natural linear retention spaces. An analysis of the natural retention areas historically present along the Bavarian Danube shows that relining retention areas during larger flood events hardly has an effect on flood water wave height. Dyke relocations yield a lower level locally (broader flow) and a certain delay in the discharge wave when there are several of them (see Section 4.3.6).
- Cultivating river barrages. Observations on pre-sinking followed by higher backwater level in river barrages (for example, Geisling and Straubing barrages) show that the additional retention volume created by pre-sinking largely fills the impounding again, with average volume/discharge conditions (m^3 retention volume/m^3/s discharge) right when the Danube's HQ100 wave arrives. It may be possible to implement tailored management strategies at various Danube backwater levels in the event of flooding as general support. However, this is no replacement for manageable flood polders. Potential effects of pre-sinking on wave overlapping taking place downstream must be observed.
- A much higher effect can be yielded (under the same retention volume) by non-linear (i.e. with intake barrier, i.e. summer dykes) and controlled retention areas in return, as they influence flood peaks in a targeted manner. An individual effect analysis at the 12 potential flood polder sites reveals that each site can make a significant contribution of approximately 30 km through local relief (peak reduction).
- Positive effects were evidenced on a trans-regional scale as well. The extent and range of the respective apex reduction are of course dependent upon multiple general conditions, including the usable retention volume, site location, and interaction between river and floodplain areas. However, the respective hydrological scenarios involving location and duration of the rains causing the flooding and the interplay among multiple relief processes are highly decisive for the trans-regional effect and vary highly in their degree of effect. The effects matrix is complicated at very the least, if not complex in a non-linear manner.

This means it will not be sufficient to create larger manageable retention options in just one Danube section in order to react flexibly to any respective flooding situation. Larger controllable retention spaces should be created in each related hydrological river section, i.e. between the estuaries of the tributaries affecting river discharge. A combination of controlled flood polders in one respective Danube section (about 50–60 km) has proved to be particularly effective. The reducing of flood wave peak(s) has to be repeated on a regular basis along its course, in order to attain a trans-regional effect.

Volume observations have clearly shown that the minimum volume required to attain a certain capping percentage in wave peaks increases significantly from section to section along the course of the

Danube. In the flood events observed, for example, attaining a reduction in the discharge peak of 10% in the Danube section between the Iller and the Lech estuaries (reference level: Donauwörth) required at least 10 million m^3, in the Danube section between Lech and Naab/Regen estuary (reference level: Ingolstadt) at least 20 million m^3, and in the section between the Regen and Isar estuaries (reference level: Schwabelweis) over 40 million m^3. This represents the minimum volume needed for optimal capping. As regards interaction with other retention effects, such as the marsh upstream from Donauwörth, significantly higher volumes may be necessary depending on the flood polder site.

The prioritisation of the individual sites based on the flood polder effect and the illumination of additional aspects provide decision-making assistance that can be used to select the flood polder sites in the next planning step. The flood polder site with top priority in the two Danube sections named above bears a high level of conflict potential in other fields, especially nature conservation. This is why it is recommended to include an additional alternative site in further planning steps.

Based on the findings from this study (Rutschmann, 2013), it is recommended to proceed as follows:

- Danube section Iller to Lech estuary. In this section, the flood polder sites Höchstädt and Schwenningen rank best in prioritisation. Feasibility studies should be contracted for these sites. They are both in the vicinity of the high population area, Donauwörth, and can be controlled even when there is extreme Lech influx when needed. At least one of these two flood polders should be implemented over the medium-term. Upstream, a marsh also quasi represents a flow polder that activates additional retention volume under extreme discharge due to increased channelling out.
- Lech to Naab/Regen estuary. The construction of the flood polder in Riedensheim was due to commence in early 2015. Additionally, feasibility studies will be performed for the Bertoldsheim flood polder site (performs best in the prioritisation) as well as for the Großmehring site (location similar to Katzau, for which a study is already available; see below). Over the medium term, at least one further flood polder should be built in addition to Riedensheim.
- Regen to Isar estuary. In order to attain an effect in this section similar to the upstream Danube sections, the flood polders of Wörthhof and Eltheim will also need to be implemented in addition to the flood polder Öberauer Schleife, which is already in planning. Hence, a feasibility study should be commissioned for both sites.

All sites suggested for future polder projects should be legally secured in a timely manner. It is also recommended to secure the remaining three sites in the most upstream Danube section as well in order to keep options open for the future.

The suggested controlled flood polders (at least one in the Iller to Lech zone, at least two in the Lech to Naab/Regen zone and all three in the Regen to Isar zone) will make it possible to establish effective elements for flood management delivering significant relief for the existing flood protection installations along the Danube, thereby reducing the residual risk. Alongside the other elements, e.g. overflow sections and erosion-resistant dykes, they should be viewed as a component for attaining resilient safety systems along the Bavarian Danube.

4.3.6. Wider verses higher. Political discourse often addresses the assumption that 'natural' measures, especially natural river retention, can be used to hinder flooding, or conversely that expanding the river has increased the size of flood waves. This is only true in part, as the derivations in the following will illustrate: until around 1800, the Danube flowed in constantly changing curves and streams in a riverbed that was kilometres in width (Figure 5).

This is why a channelisation was performed on the Bavarian Danube between 1806 and 1867. The corrective measures lowered the river as planned. Additionally, at the start of the 20th century, dykes were erected along the river. The flooding that used to take place throughout the entire Danube valley now only occurs to an extremely limited extent. In order to stop the continued lowering of the river bed, barrages were erected on the Bavarian Danube, which also served the purpose of energy generation. Several barrages are designed such that natural inundation can still take place when water levels are high.

Despite this, huge retention space has been lost along the Danube due to the river channelisation and the barrages. Today between Ulm and Straubing, around 330 million m^3 less retention volume is activated than was the case 200 years ago during an HQ100 event.

In order to evaluate the effects the river channelisation has had on the spread of flood waves along the Danube, the Bavarian Danube Valley was modelled in its historical state in order to research the effects the river correction had on the spread of high water waves. To that end, a land model was produced with the help of historical maps to reproduce the Danube from the year 1800. With the help of two-dimensional hydro-dynamic numerical simulation, various flood events were computed and the spread of high water waves was evaluated.

The studies of the historical Danube show that nearly the entire Danube Valley was available as a floodplain/retention area prior to the channelisation. The result from the calculations (base: Pentecost flood 1999) shows that the retention volume between New-Ulm and Straubing reduced by approximately 330 million m^3 compared to the historical values; however, this has had hardly any influence on the height of the flood peak, which is remarkable. Compared with the historical scenarios, however, it can be demonstrated that today's status yields a significant accelerated flood wave on the Danube, which can have effects on overlapping with flood waves from the Danube tributaries. Historically smaller events were also better dampened.

Based on the knowledge available to us today, we assume the following general properties: in larger water bodies, the dyke relocations largely reduce the local water level by removing bottlenecks and

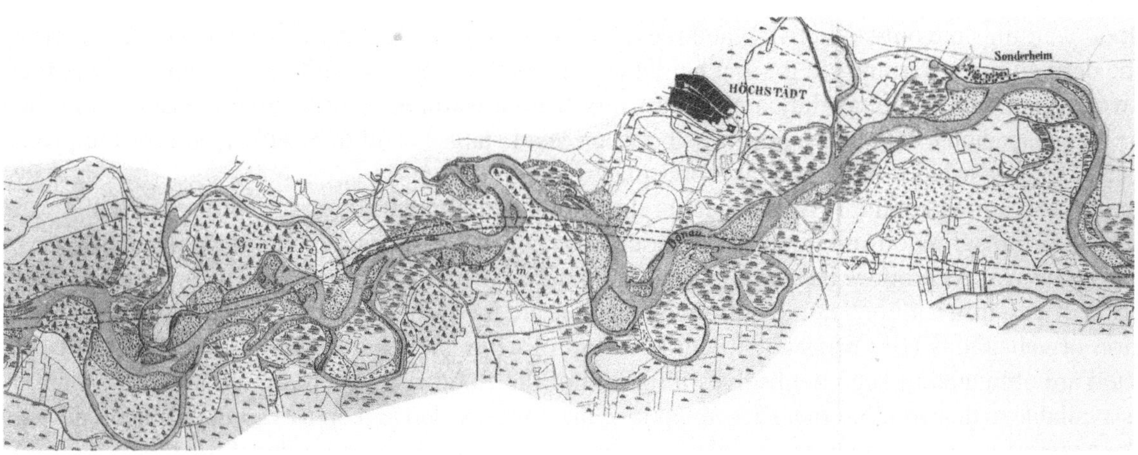

Fig. 5. Danube correction – channels planned at the start of the 19th century.

enlarging cross-sections for discharge. However, no nominal effects on an HQ100 peak are yielded down-river. Newly acquired retention space is usually inundated even in the event of smaller events and/or as flood water rises, and is therefore hardly available any more for capping and/or retaining high water peaks. In smaller rivers, these kinds of dyke relocations can also represent a component to preventative flood protection with effects on the height of the flood peak. However, the effect drops as the intensity of the flood event rises. The ratio between the volume of the flood wave and the activated retention volume is decisive for the effect on discharge reduction. The volume of the flood wave increases along with the size of the catchment area. Thus, the positive effects of dyke relocations (reduction in flood peak) decrease as the size of the catchment area increases. Thus, HQ100 protection according to plan cannot be implemented by solely relocating dykes in upstream river sections.

This is why the immediate reasons for dyke relocations are usually ecological aspects, such as targeted water meadow cultivation and synergies with nature conservation aims.

Irrespective of this, it is urgently recommended – given our anthropogenically moulded system – to retain all inundation zones (maintenance of retention areas) and relief modifications (this also includes the drying of wet zones) still available, and to reduce or stop land utilisation.

4.3.7. Flood prognosis and forecast. Timely warning of flood events avoids damage and saves lives. Since 1890, a flood warning service (HND, *Hochwassernachrichtendienst*) has worked to avert flooding and ice hazards in rivers and levels in Bavaria, as laid out in flood notification plans. The core tasks of 'measure', 'report' and 'warn' have remained unchanged to date. However, today's technical capabilities are revolutionary compared to those in the beginning. Our information age expects reliable information to be available at all times, everywhere and via all media. The HND in Bavaria endeavours to meet these requirements to the best extent possible, receiving support from an optimally structured water management system to that end. Based on recent experiences, the HND was overhauled and has the following status (in 2014):

The Flood Notification Centre (HNZ, *Hochwassernachrichtenzentrale*) creates a status report for all of Bavaria, and makes this information available on the HND's Internet site. Five regional Flood Forecast Centres (HVZ, *Hochwasservorhersagezentralen*) are each responsible for a large river zone, producing forecasts and providing warnings to the Water Management Agencies (WWA, *Wasserwirtschaftsämtern*), which have 17 office sites.

Flood warnings are only effective when they are met with prepared measure plans and those measures can be derived from warnings (e.g. flow height at gauge). Community notification plans are the focus of this work again and again, as they are the last link from a warning to protective measures and the citizens. They especially lay out relationships between level statuses and their effects and/or the protective measures to be carried out, as well as flood plans.

The flood warnings of the 17 WWAs are broken down into three warning levels: pre-warning, warning against bank transgression and flooding of agricultural zones, and warning against flooding of zones with edifices. The WWAs produce the warnings and forward them to the notification offices linked to the HND (district offices, urban municipalities), from where they are forwarded to the communities and disaster prevention organisations (fire brigades, integrated control centres). After receiving a warning, the parties affected are obligated to keep themselves informed of the flood risk. There are various information services available to that end (Internet sites, telephone messages, video text on Bavarian public broadcasting).

The trigger for the warnings can be observed transgressions of a warning level for reporting, or a pre-warning issued by the HVZ and the HNZ. The decentralised warning system accounts for the local circumstances in the WWA zones and provides for on-site presence. The provision of flood information and warnings in video text, a measurement announcement service and the Internet means the authorities and the public are informed simultaneously.

The foundation for the HND consists of measured data, especially the gauges in the rivers. A total of 320 gauges are specifically used for flood warning, another 480 gauges and some 850 precipitation measurement points are used for forecast calculation via surface runoff models, as well as for other reasons. Measures for improving resilience have been undertaken, such as reinforcing the flood stability of measurement equipment beyond HQ100; data channels and equipment at the most important gauges were doubled in order to avoid outages. The data are automated centrally at a database and merged every 15 min around the clock. There are cooperation agreements with the weather services and flood warning services in neighbouring countries, and with hydro-electric power plant operators as well as with other organisations via automated electronic data exchange. The forecasts are also exchanged. External forecasts are fed into local models.

In this manner, automated hourly forecasts for the next 48 h are computed for the River Inn from the Swiss border to Passau, and on the Danube from Regensburg to Vienna through the neighbouring states of Tyrol, Salzburg, Upper Austria, Lower Austria and Bavaria, and then published on the Internet along with margins of uncertainty. The central data store means the integrity of the data can be secured. Measured data and forecasts come from one source. Their availability is secured through redundant data storage via the replication of the master database at the geographically separated flood forecast centres.

The official gauge measurement network for larger rivers encompasses around 50% of all communities, which are connected to the HND via flood notification plans. The special Storm Weather Plan put forward by the State Ministry of the Interior, in contrast, implements all communities in Bavaria; it governs the forwarding of storm warnings from the German Weather Service (DWD, *Deutscher Wetterdienst*) to the communities. Among other things, this makes it possible to send warnings in a short-term, highly effective manner regarding heavy rain events, which generally affect local and especially smaller rivers. The storm warnings from the DWD are integrated into the HND Internet site.

Within the scope of implementing the EU Floods Directive (2007), warnings are also issued for medium likelihood of flooding (at least a 1 in 100 year event), extreme events and low likelihood events, with corresponding measure plans created by the communities.

4.3.8. Heavy rain and flash floods insurance. The large floods that raise our interest in trans-regional media are usually river floods. We have a relatively high level of knowledge about these scenarios. We have prediction models, gauges and measure systems; we know the flood zones and have leveraged numeric simulations to work out all feasible hydraulic parameters, such as flow velocity, flow depth, shear stress, etc. In addition to these river floods, however, there is a flood risk that receives much less attention in the media, known as urban flash floods. What is fatal about these events is that they can arise practically anywhere, even far away from bodies of water, there are hardly any reliable options for prediction (a rain forecast, at most) and nominal pre-warning times, and they will also occur with increasingly frequency in the future as a result of climate change. They are often created by local occurrence of heavy summer rain events, at times in combination with storm and hail. These events frequently far surpass the capacities of property drainage systems, waste water systems and smaller bodies of water. Urban drainage systems are not usually big enough to account for these rare events. The strategy of measuring a rain water drainage system for a 100-year event is presumably doomed to fail. Here, new ways of channelling and retaining must be found, and the idea of private object protection must be moved more into the foreground. It is often the case that raised cellar shafts and slightly heightened door sills are sufficient to prevent the greatest damage.

According to the many years of experience of German insurance providers, half of all standard flood damage results from these kinds of localised extreme events, urban flash flooding, and is usually far from actual flowing bodies of water (DWA, 2013). This means that these flash flood events are in the same league in terms of money as the flooding of big rivers so often seen in the media.

A natural hazard insurance policy is well suited to these events, which can occur anywhere, particularly as the problem of anti-selection is not an issue here. Anti-selection means that usually only those building owners who live next to a river and experience frequent flooding take out such an insurance policy. However, if that is the only group of persons who purchase insurance, then the charge for an insurance policy would be prohibitively expensive. Purchase of a natural hazard insurance policy is voluntary in Germany. The coverage sum is largely dependent upon the property's risk level. To that end, the German insurance industry has developed a zoning system with four hazard classes. The data from water management offices, such as computed flood zones, are taken into account here. The risk in the lowest hazard level is largely determined by heavy rain events, and insurance policies are correspondingly financially affordable there. Currently however, the insurance rate in Bavaria is only approximately 21%, while it is approximately 33% in Germany as a whole.

However, it is here in particular that a social solidarity is required. State and community offices use public funds to erect suitable flood protection against high river water for regions with human settlements along rivers. However, the level of protection is usually restricted to a 100-year event (e.g. dykes and technical flood protection systems are designed for a 100-year event). State flood protection measures, however, reduce the hazard classification and the purchase of a natural hazard insurance policy becomes affordable for everyone. With an insurance policy of this kind, a private citizen hedges the residual risk, specifically the potential of failure or overload in state flood protection installations. Simultaneously, this insurance policy also covers the residual risk of flooding due to urban flash flooding. The combination of a lower risk of high river water through state flood protection installations and comprehensive insurance coverage as regards urban flash floods makes natural hazard insurance accessible for the public at large.

4.4. The burden of maintenance

4.4.1. Flood protection infrastructure: moving toward safety and prosperity, or running in circles?

Valleys, rivers and river forelands were the seeds for the historical development of human settlement and the starting point of today's societies. Goods transport, water utilisation and transportation channels are all factors which led to human settlements arising in zones at risk of flooding. However, this danger was accepted, as the location also yielded an economic upswing, which led to increased prosperity. While people used to be obligated to live with flooding, capacities then emerged for correcting river runs and erecting the first flood protection installations at known focal points. This reduction in flood frequencies automatically increased the utilisation of river zones (with consequent damage potential) while the awareness for flooding hazard simultaneously dropped. However, many of these flood protection installations were never intended to protect the damage potential of modern human settlements and cities, but were rather designed to help prevent famine in the Danube region and to protect harvests against flooding, which occurred rather frequently. These installations were able to withstand flooding events for a long time. The structural change that incurred (away from agriculture), growing prosperity, the enormous boom in construction and the growth in population, however, have led to a situation bearing enormous damage potential. For example, between 1840 and 2012, the population in Bavaria grew from 3.8 million to over 12 million residents. The flood protection installations necessarily have to be raised and reinforced in order to meet today's protection level for settlement areas (HQ100). However, this retrofitting of installations automatically leads to an increase in damage potential, as there are no longer construction restrictions in these HQ100-protected zones. The damage potential rises again. Does that mean there is a need for more protection as well? It is especially when new flood protection measures are built that, in addition to the technical protection measures, further elements of flood risk management must be implemented. This is the only way to break the spiral of more and more need for protection. Even when doing this is successful, at the least a dependency on these (new) protection installations is yielded, which can no longer be reversed, as relief and settlement structures develop in an irreversible manner. For these societies, this means that each year a certain budget has to be provided to maintain the flood protection equipment that has become necessary to the people living there. Otherwise, there is a risk of collapse. Once the trigger for economic development, flood installations can become an ongoing burden. Whether they are streets or dykes, this maintenance has been consuming more funding than is available for new construction.

In summary, one must be sure that the installation of each new flood protection measure is indeed desirable today, and aware that they will quickly become essential for the region following construction, and that there will be no going back.

4.4.2. Example: resettlement. While this is only possible in individual cases, there is indeed a rising awareness for conscious decisions to no longer continue expanding flood protection in certain regions that are already settled. The experiences from the flooding in 2013, however, have demonstrated that financial decisions cannot be the sole factor here. Human suffering, trauma and psychological stress with later effects on health must also be accounted for. A typical approach here in Bavaria is to at least provide financial support for voluntary resettling to the residents in zones that cannot be permanently protected with flood protection in line with state of the art technology.

4.4.3. Example: development of maintenance strategies in torrent control in the Alpine region. The current situation in torrent control in Bavaria's Alpine region can be described briefly as follows: over the past 100 years, some 50,000 protective structures have been built along torrents and their drainage areas. Many of these structures are no longer in line with modern requirements. Their maintenance and restoration consume significant levels of resources; it is a Sisyphean task. Year on year, attempts are made to restore the existing structures with existing capacities, and to renew them as they collectively decay (Figure 6). Moreover, there is even more need for new expansion and protection measures in order to protect new and more sensitive infrastructure and to stay on top of the rising requirements from climate change and society's vulnerability. In line with the discussion of sustainable development earlier in this paper, this yields a general specification for technical development (Grambow, 2012): 'On both the level of the individual installations and on the level of entire technological systems, there should be permanent monitoring and enhancement to the technology in the direction of sustainability criteria.' General life experience reinforced academically by resilience theory, however, shows us how difficult it is to depart from a path that has already been started and solutions that are well established. There is

Fig. 6. Just one of thousands of torrent control structures that require maintenance.

danger that 'due to holding on to traditional solutions, funds are being 'sunk' and the conservation phase of a system is being stretched so much that there is no homogeneous new development loop on the other end, but rather a collapse similar to a revolution in nature' (Grambow, 2012).

Applied to the situation described above in torrent maintenance, the established solution represents the 1-for-1 restoration of structures (i.e. restoration or replacement at the same location and in the same construction model), without checking whether optimisation is an option. A crisis-like collapse then occurs when the protection systems thought to be safe gradually fail due to maintenance no longer being financially feasible. Against this background, a comprehensive research project was launched by the Bavarian water management administration with the objective of finding a way to optimise the protection systems with regard to minimisation of maintenance expenditure, and modification in framework conditions as well as long-term adjustments and further expansion requirements. In principle, there is a similar situation with regard to first-order and second-order rivers. This can presumably be applied to many further fields involving infrastructure installations (e.g. roads) in a similar manner. Once installed, our society becomes more and more dependent upon these installations. Once considered to be a 'relief' for society, for future generations it quickly becomes a dependency on 'life-sustaining systems' that can no longer be reversed, and therefore turns into a significant burden on future budgets over the long term.

4.4.4. Reconsidering past solutions with regard to changes in general conditions. While dyke systems erected at the start of the last century were frequently intended to protect agricultural spaces and for sustaining food production (degree of protection ~ HQ30), these systems have to protect settlements and industrial zones that have grown since that time, which means they have to meet an entirely different objective (degree of protection HQ100). The changed requirements and general conditions to which torrent control has been subjected over the past decades are much more versatile today. Settlements have expanded to the scree at valley exits. The forest and vegetation situation as well as uses (e.g. mountain restaurants, ski slopes, forestry, tourism) have changed over the decades. In some places, reforestation is leading to stabilisation in mountain faces, and thus to a reduction in discharge and bedload in torrents. Short-term individual events quickly have the opposite effect, such as storm damage and avalanches. Among other things, new construction methods have developed due to changes in machinery, or were even necessary due to changes in personnel. Modern computation models permit precision in process assessment and dimensioning in protection systems. Databases and the documentation of past events also facilitate re-assessment of the current situation. Added to this are new legislative requirements in nature conservation and the two EU Directives (Water Framework Directive, 2000; Floods Directive, 2007). The knowledge that a residual risk remains for all protective measures and that our financial and personnel resources are limited requires the continuous forcing of transition away from pure hazard averting to a sustainable risk culture in natural hazard management. The list is merely an outline, yet it shows that costly 1-for-1 restoration and holding on to what are frequently historical solutions should at least be questioned and optimised (Figure 7). In this context, however, optimisation must not be placed on the same level as an increase in efficiency, which is often misunderstood. An increase is efficiency solely targeted at trimming a protective system for one reason and to prepare for a design situation in a cost optimised manner

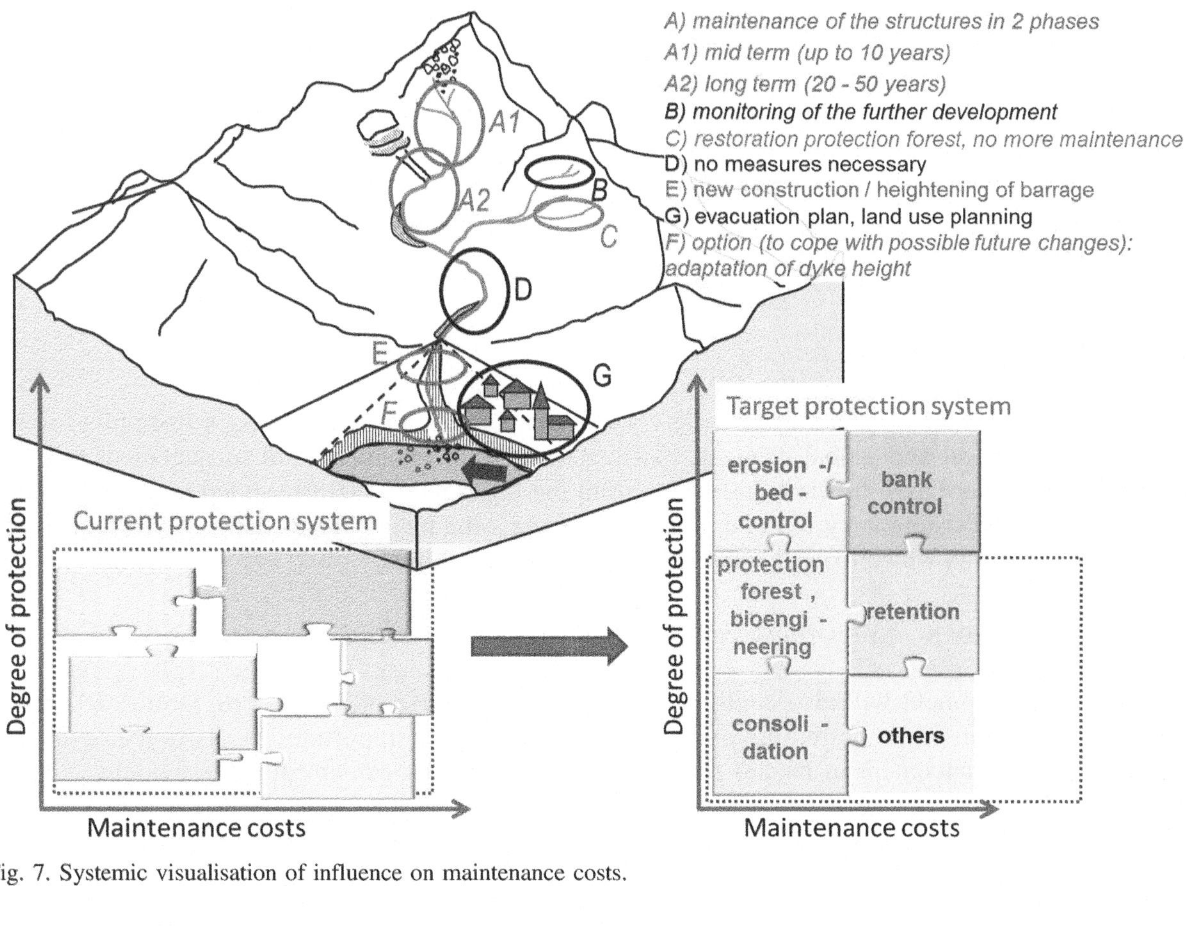

Fig. 7. Systemic visualisation of influence on maintenance costs.

which does not take critical overload properties and 'hard' system tipping points into account can never be sustainable. Optimisation of protection must consider the aggregate risk as well as costs, and lead to compliant, resilient and adaptable systems.

As an outcome from a research project on four torrent zones featuring well over 100 individual structures, research was performed regarding optimisation of maintenance and future expansion of reconstruction of the protection system in pilot studies. The objective was to derive a guide for the future maintenance strategies in the torrent catchment areas in Bavaria. The intention was not to work on specialised topics like transportation processes, driftwood, hydrology, damage analysis, monetary damage and efficiency assessment, risk analysis and assessment, geology, and hydraulics in line with state of the art technology, but rather to illuminate an integral and balanced aggregate observation that can be applied in practice.

Structures were identified in all drainage zones that are no longer needed in their respective state from a modern point of view. Simultaneously, however, weak points in the protective systems and additional need for expansion were revealed in all the studies. A 1-for-1 maintenance was therefore not the most sustainable solution in all catchment areas. There were also intensive efficiency considerations regarding the maintenance cycle for key and standard structures with simultaneous consideration of the state a structure is in (decay over time), as well as at times comprehensive drainage zone observations of historical states and even of future scenarios, like changes in wildlife and forest structures.

The implementation of this strategy must of course be accompanied in practice by intensive communication. Moreover, a sustainable master concept can only be attained with the best possible utilisation of knowledge and experiences on site.

Part of this is supplementing the process of technical rethinking with risk management measures on the demand side, i.e. via risk communication and passive and active prevention on the infrastructure and settlement side (warning and deployment plans, marking of hazard zones, restrictions in urban land-use planning). This is how modern participatory methods will become elementary components to the future torrent strategy.

5. Conclusion

Flood protection is an integral part of society's development. To meet a growing vulnerability in the context of geopolitical and global changes, existing systems of natural hazard management must be reviewed on a regular basis. In Bavaria, lessons from the big flood in 2013 have led to a fundamental modification of the existing integral flood protection strategy, the main characteristic of which concerns the handling of events which exceed standard structural design limits.

5.1. The systemic core to this approach consists of holistic scenarios and a targeted resilience strategy

Technical development will also continue throughout the course of Action Programme 2020plus, requiring adjustments in programming and even opening up new opportunities. Its 6-year cycle for adjustment and advancement in hazard maps and flood risk management plans is evidence of this fact.

Irrespective of state guarantees stipulated by society (including communal self-responsibility), sustainable and comprehensive protection can only be attained with collaboration from everyone. Here, dialogue regarding risk for the revising of flood risk management plans will attain particular significance.

In order to continue the high quality planning, building and safe operation of effective flood protection installations in Bavaria in the future, it will also be necessary to have well trained, responsible technicians and engineers in planning agencies, construction companies and public administration. The Bavarian water management administration has committed itself to these tasks and will meet this obligation with appropriate equipment and personnel.

References

DWA (Deutsche Vereinigung für Wasserwirtschaft, Abwasser und Abfall) (2013). *Starkregen und urbane Sturzfluten – Praxisleitfaden zur Überflutungsvorsorge – T1/2013.* DWA, Hennef, Germany.

EU Water Framework Directive (2000). Directive 2000/60/EC of the European Parliament and of the Council of 23 October 2000 establishing a framework for Community action in the field of water policy. EU, Brussels.

EU Floods Directive (2007). Directive 2007/60/EC of the European Parliament and of the Council of 23 October 2007 on the assessment and management of flood risks. EU, Brussels.

Grambow, M. (ed.) (2012). *Nachhaltige Wasserbewirtschaftung.* Springer Vieweg, Wiesbaden.

MunichRe (2014). *Topics: Geo natural catastrophes 2013.* MunichRe, Germany. http://www.munichre.com/en/reinsurance/magazine/publications/index.html.

Rutschmann, P. (2013). *Verzögerung und Abschätzung von Hochwasserwellen entlang der bayerischen Donau.* Technical report of the Lehrstuhl und Versuchsanstalt für Wasserbau und Wasserwirtschaft der TU München. https://www.wb.bgu.tum.de/.

Walker, B. & Salt, D. (2006). *Resilience Thinking.* Island Press, Washington, DC.

Wiebeking, C. F. V. (1817). *Theoretisch-practische Wasserbaukunst.* München (1811–1817).

Losing memory – the risk of a major flood in the Paris region: Improving prevention policies

Charles Baubion

Organisation for Economic Co-operation and Development (OECD)
E-mail: Charles.Baubion@oecd.org

Abstract

This paper provides a snapshot of the key findings of an Organisation for Economic Co-operation and Development (OECD) review of flood risk prevention policies in the Paris metropolitan area. With an innovative flood risk assessment, the study shows that a major flooding of the Seine River, similar to the flood disaster of 1910, could affect up to 5 million residents in the greater Paris area and cause up to 30 billion (10^9) euros worth of damage. Economic growth, jobs and public finances could also be significantly affected at the national level. The OECD Review on Flood Risk Management of the Seine River – commissioned by the basin organisation Seine Grands Lacs with the French Ministry of Ecology and Île-de-France regional council – recommends raising risk awareness among citizens and businesses, and improving the resilience of the metropolitan area to flood risks. Recent floods in Europe and New York City's Hurricane Sandy disaster in 2012 illustrated the vulnerability of today's ever-denser cities to flooding and the need to adapt critical infrastructure systems to be able to cope with extreme weather events. The OECD review suggests ways to minimise the risks and better prepare the Île-de-France region. It notes that proposed projects to develop and expand the city's transport and logistics networks offer an opportunity to put some of its suggestions into practice.

Keywords: Critical infrastructures; Flood; Risk governance; Risk modelling; Urban resilience

1. Introduction

One of government's key responsibilities is to ensure that large metropolitan areas are resilient to major risks, to guarantee the safety and welfare of the public, and maintain public trust. Ministers from the Organisation for Economic Co-operation and Development (OECD) recently adopted the Recommendation on the Governance of Critical Risks to insist on this responsibility (OECD, 2014b). In the light of the Great East Japan Earthquake or the flooding associated with Hurricane Sandy in New York,

---

This is an adapted version of OECD (2014), 'Executive summary', in OECD, *Seine Basin, Île-de-France, 2014: Resilience to Major Floods*, OECD Publishing. http://dx.doi.org/10.1787/9789264208728-3-en.

doi: 10.2166/wp.2015.008

governments, local authorities and civil society have become increasingly aware of the fragility of major urban centres when disasters occur and of the degree to which critical systems are interconnected. Within the framework of global debate on climate change, it is important to question the capacity of cities to adapt to extreme weather, water or climate events. Flooding in Paris on the scale seen in the historic flood of 1910 could constitute a major shock and pose an unprecedented challenge to public policy. The Greater Paris urban development project provides a unique opportunity to promote good practices in resilience to encourage sustainable development.

This paper presents the key findings of the OECD Review on Flood Risk Management of the Seine River in the Île-de-France region, conducted as part of the OECD High Level Risk Forum (HLRF) in the OECD Directorate for Public Governance and Territorial Development. The Seine flood risk review is part of an OECD series examining risk management policies, and follows on from surveys conducted in Mexico, Italy and Japan. It offers an evidence-based analysis and public policy recommendations to enhance prevention of the risk of River Seine flooding in Île-de-France, developed through an inclusive and open dialogue with all stakeholders in Île-de-France.

As a basis to the analysis, an innovative flood risk model was developed to assess the impacts of different flood risk scenarios. The impact of such an event is assessed in relation to public well-being, the continuity of operation of institutions and businesses, and the economy. The economic impact could be significant, including at national level, because of the importance of the Île-de-France region to French gross domestic product (GDP). The exploratory assessment of the macroeconomic impact of various flooding scenarios takes account of the direct and indirect effects of a shock of this nature on the national growth trajectory, employment and public finances. The findings of the review are that a major flooding of the Seine River in Île-de-France could affect up to 5 million residents, particularly as a result of the effects on critical infrastructure networks. The economic damage has been estimated at between 3 and 30 billion[1] euros for direct damage together with a significant macroeconomic impact at the national level.

The report also notes the progress made and the many policy tools employed in flood risk prevention in France that could contribute to best practices, and highlights that they have not significantly contributed to increase the resilience of the Paris region to major floods. It highlights how the complex governance of this strategic region has limited the effective implementation of prevention policies. Today there is a dynamic in progress to improve governance. Long-term investment projects are envisaged for the region under the Greater Paris project. A transparent, accepted approach to the question of risks will help draw maximum advantage from the opportunities that arise. The review offers policy-making guidelines for the future to enhance flood-risk prevention in Île-de-France: development of a risk culture among the general public, companies and decision-makers; innovation in urban areas to foster resilience; and promoting a consistent financial approach based on economic efficiency, consideration to the long term and equity in funding.

2. Impacts of a major flood of the Seine River in the Paris-Île-de-France region

2.1. A historic reference: the great 100-year flood of 1910

While the possibility of a major flood of the Seine River may initially seem remote, it comes back regularly and arouses public attention, as was the case during the spring of 2013 when floods took

[1] Billion = 10^9.

place upstream of the Seine River basin. Even though the flooding did not cause any major damage, it revived the issue of risk management and the region's vulnerability to flooding. The prospect of a historic event is a key concern for French risk management stakeholders.

One of the features of the flow-rate of the Seine is its interannual variability, characterised by extensive floods. The 1910 flood is the best known with an estimated peak flow-rate of between 2,400 and 2,650 m^3/s in Paris, more than eight times the river's average. This well-documented flood acts as a benchmark. Its return period is estimated at one hundred years. The flood reached 8.62 m on the Austerlitz gauge. The 1910 flood was particularly destructive in the context of an era marked by industrial and technological progress (Caisse Centrale de Réassurance, 2013). Such events illustrate the difficulties societies have in striking a compromise between economic development and managing the increased vulnerabilities of society and multiple economic sectors.

Over the past four centuries – the Seine flood records date back to 1649 – an exceptional flood level has been exceeded around four times per century (a flood is deemed exceptional when it exceeds 7 m on the gauge) and the worst flood was 8.96 m in 1658. Although no major flood has occurred for almost 60 years, the twentieth century featured major 50-year floods in 1924 and 1955. The last major flood was in 1982, and its return period was 10 years. 1924 and 1955 also saw major flood events in the Paris region and in the entire Seine basin. Nevertheless, the lack of a significant flood for more than 60 years tends to lessen the memory of risk (Etablissement Public Territorial de Bassin (EPTB) Seine Grands Lacs, 2013). Seine floods are characterised by their slow progression and, as a result, the period of submersion may be very long. For instance, the waters took almost 2 months to subside in 1910.

Even if the effect of climate change on the frequency and extent of the Seine floods is still uncertain (Ducharne *et al.*, 2011), greater floods than that of 1910 are still possible, such as the one that occurred in 1658 – the 17th century also saw three major floods in nine consecutive years. In other countries, many recent floods have significantly exceeded 100-year levels. This was the case with the floods in Queensland, Bangkok and Pakistan, as well as during the coastal flooding following hurricane Sandy in New York and the 2013 floods in Germany (Table 1). The 2007 EU Floods Directive uses the 1,000-year frequency as a reference for extreme events.

Table 1. Return period for recent major floods.

Country or city	Year	Flooded river	Return period
Prague	2002	Vltava	500 years[1]
United Kingdom	2007	Severn	200 years[2]
Pakistan	2010	Indus	>100 years[3]
Brisbane	2011	Brisbane	120 years[4]
Bangkok	2011	Chao Phraya	> 100 years[5]
New York	2012	Floods linked to Sandy*	400–800 years[6]

*Floods linked to Sandy are different as they were caused by sea storm surge.
[1]Zizkova (2004).
[2]Severn Trent Water (2010).
[3]Japan International Cooperation Agency (2012).
[4]Queensland Flood Commission Inquiry (2012).
[5]Aon Benfield (2012).
[6]Lin *et al.* (2012).

2.2. *Despite investments in protection, increasing urban development and the interdependence of critical infrastructures have accentuated vulnerability*

Since 1910, the risk of a flood of the Seine River in the Île-de-France region has been reduced in various stages by protective structures, including dams built upstream and river development starting in the 1920s, then continued from the 1950s up until the early 1990s. Major investments have nevertheless been limited in recent decades and it appears that protection levels are not up to the standards of many other comparable OECD countries, particularly in Europe (Table 2). Furthermore, exposure to risk and the resulting vulnerability have been accentuated by increasing urban density in this leading economic area of France (Figure 1), as well as by the construction of a large number of industrial areas and critical infrastructures (transport, energy, communications, water) along the Seine River. The interdependence of these networks, the interpenetration of production lines and their 'just-in-time' operation, the key role played by the mobility of people and exchange in a dynamic economy, urban development, and the concentration of populations and capital are just a few of the many factors which increase modern societies' vulnerability to shocks (OECD, 2011).

Table 2. Levels of protection in major cities in the OECD.

City	Protection level	Protected area	Date of implementation
Paris urban area	100 years	Paris	Existing
	30–50 years	*Départements* in the inner Paris suburbs	
Frankfurt[1]	200 years	New districts & critical infrastructures	Existing
	100 years	Urban area	
London[2]	1,000 years	Coastal areas	Existing
	10,000 years		Objective 2100
New York[3]	100 years	Coastal areas	Project 2020
The Netherlands[4]	1,250 years	River areas	Existing
	4,000 years	Coastal areas with small populations	Existing
	10,000 years	Areas below sea level	Existing
Oslo[5]	100 years	Residential buildings	National standard
	200 years	Industry and critical infrastructures	
Tokyo[6]	200 years	River and coastal areas	Current project
Köln[7]	100 years	River areas	Existing
	200 years	Particularly critical areas	
Vienna[8]	10,000 years	River areas	Existing
Bratislava[8]	1,000 years	River areas	Existing

[1]Land of Hesse (2007).
[2]Environment Agency (2012).
[3]New York City Special Initiative for Rebuilding and Resiliency (2013).
[4]Netherlands Ministry of Infrastructure and the Environment (2012).
[5]Norwegian Directorate of Water and Energy Resources (2009).
[6]Cooper & Matsuda (2013).
[7]World Bank (2012b).
[8]Kryžanowski *et al.* (2013).

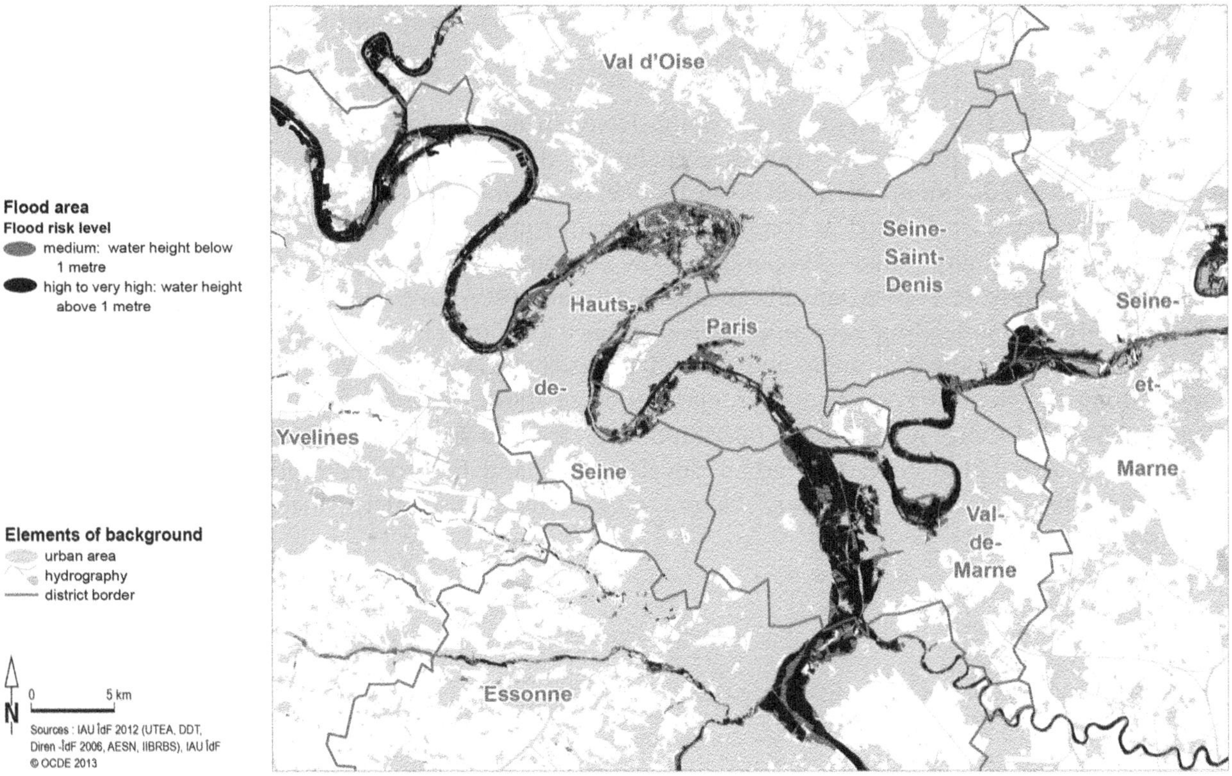

Fig. 1. Map of the floodplain for a 100-year flood.

2.3. Today a major Seine flood would have important potential impacts on well-being, and on the activities of the government and businesses

The Île-de-France region represents about one third of the economic activity in France, the second largest economy in the euro area. The seat of the government and major industries are located there, as well as the main decision-making and research centres. Île-de-France represents a large logistics hub for the whole French economy (Chambre de Commerce et d'Industrie Paris Île-de-France, 2013). Interviews with critical infrastructure operators and emergency managers revealed that, in the most extreme case, a Seine flood in the Île-de-France region could have direct and indirect impacts on almost 5 million people and a large number of businesses, with significant economic, human and social effects. It could disrupt the functioning of the government and many institutions, as well as most of the infrastructures and critical networks that ensure the daily functioning of the Paris metropolitan area. Non-tangible impacts on the environment or the cultural heritage of Paris could also be severe (Ministère de l'Ecologie, du Développement durable et de l'Energie, 2011).

In total, 830,000 people living in the floodplain, or 7.2% of the population of Île-de-France, and 55,700 businesses (9.5%) providing 620,000 jobs (11.5%) would be directly affected (IAU, 2011a, b). Many official bodies, business districts and prime heritage and tourism sites are also in the flood-prone area, as are transport, electricity and water infrastructure, and several schools and healthcare establishments. The distribution of electricity could be greatly affected with almost one quarter of power sub-stations flooded or cut off as a precaution, and more than 1.5 million customers who could experience power cuts. Public transport could be affected with almost 140 km of the 250 km underground network closed as a precaution. The road network could be blocked at many points: the bridges crossing the Seine closed to traffic, due to their weakened structure, would make it impossible to travel from the right to the left bank. The drinking water supply could be interrupted around Paris where more than 5 million clients could suffer extended water cuts and 1.3 million clients see a deterioration in quality (Figure 2).

The debate on such impacts must examine the sector's interdependencies, for instance, between the critical networks (energy, communications, water, transport) and the large industrial and service sectors. A major flood could affect key economic sectors at the national level such as tourism, food distribution, or the car industry. The issues at national level are therefore major. Accordingly, preparing for a possible Seine flood (and reducing the risk) is an important, complex and sensitive issue for public policy.

2.4. Evaluating the indirect economic impacts of disaster: a literature review

In view of the substantial direct or indirect damages and losses, questions should be asked about the overall economic impact of a shock of this extent. Such an assessment needs two problems to be examined: the cascading effects of disruption to networks on companies' activities, and the macroeconomic impact at national level, given the importance of the Île-de-France region to the French economy.

Typically, the direct effects of disasters are distinguished from indirect effects (Table 3). Tangible direct impacts include damage to individuals' physical property, and to businesses (buildings, stock, tools of production) and public facilities. The methods for assessing impacts of this type in the event of a flood are well established and are detailed by the European Floodsite project (Meyer *et al.*, 2012). First, it is necessary to evaluate the spatial characteristics of the hazard (water depth, flood duration), drawing up a list of the impacts referred to above in combination with type of land use, and assessing their value. Next, the application of damage functions produces a value for damage by type of impact (as a percentage of their value) based on the flood characteristics. Modelling tools are utilised and combined with a Geographic Information System (GIS) to project hazard characteristics onto the affected area and calculate the impact depending on the assets located there.

The process of assessing companies' operating losses and other indirect effects of disasters is less obvious (Rose, 2009). Operating losses of companies affected by flood can be assessed on the basis of the direct damage they experience. However, breaks in or disruption to production processes caused by damage to companies' tools of production or stocks can have effects that cascade down the production and distribution chains, and affect the clients and suppliers of the businesses concerned. Indirect or higher order effects are those that are caused not by the disaster itself but by its consequences. They may become apparent beyond the area affected by the disaster and persist afterwards throughout the reconstruction process.

Many *ex post* evaluations ask questions about indirect impacts using surveys of businesses or microeconomic or econometric modelling. Price corrections, the effects of substitution, the stimulating effect that post-disaster reconstruction has on demand, and the Schumpeterien effect of creative destruction on productivity are all consequences that have been studied without clear conclusions being reached in every case (Przyluski & Hallegatte, 2011). It would nonetheless appear to be common ground that, where an extensive disaster affects critical networks for a lengthy period, the indirect impact is too significant to ignore (Rose *et al.*, 2012). The process of taking account *ex ante* of indirect impacts implies modelling that can incorporate the interdependence that exists between the various economic sectors and

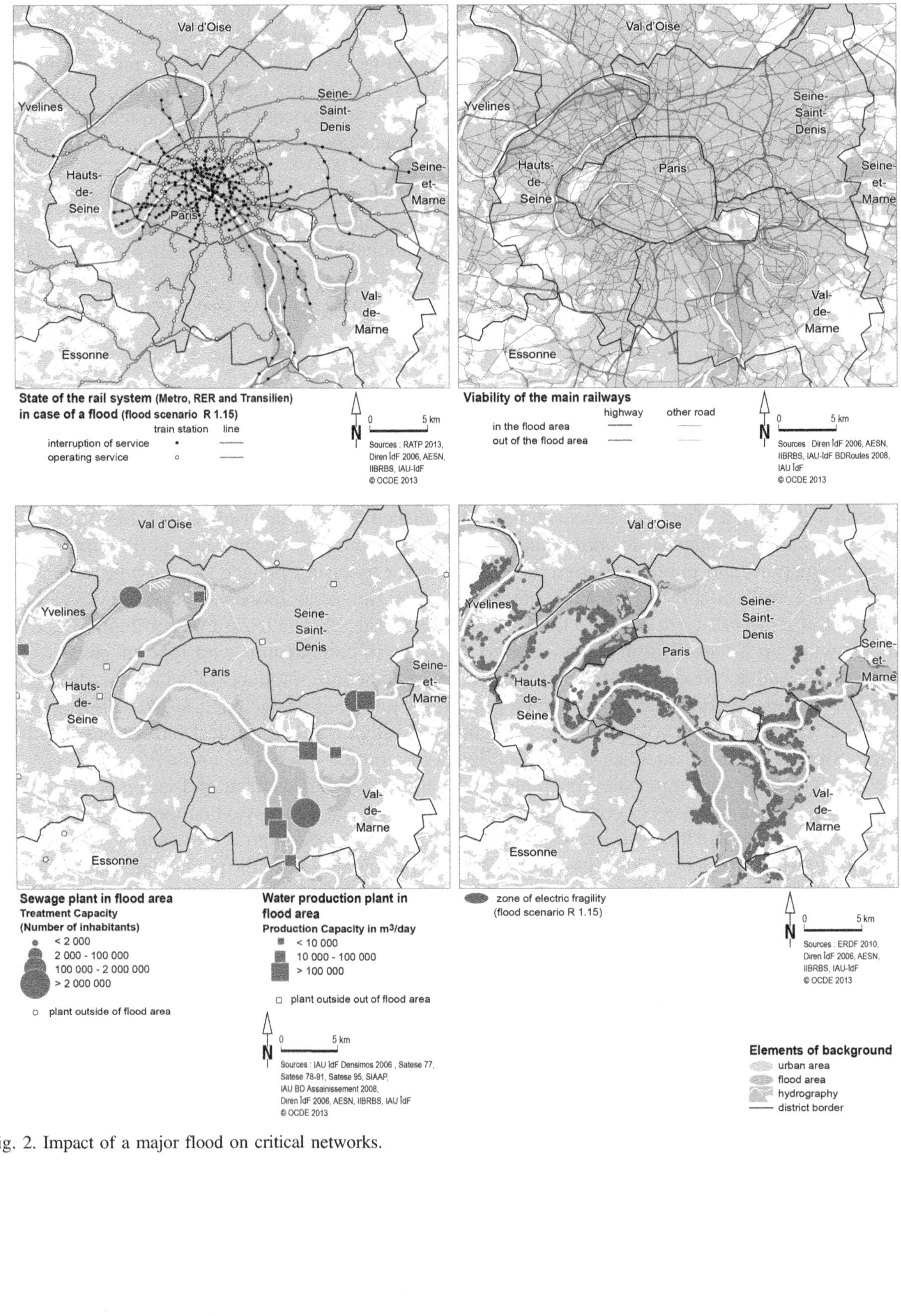

Fig. 2. Impact of a major flood on critical networks.

Table 3. Types of impact and damage.

	Tangible	Intangible
Direct impact	Physical damage to property • Buildings • Furniture • Infrastructure • Companies' stocks • Companies' equipment	• Losses of human life • Effects on health • Impacts on the environment • Impacts on the cultural heritage
Indirect impact	• Loss of industrial output • Disruption of networks • Cost of emergency response	

Source: OECD, adapted from Smith & Ward (1998); Penning-Rowsell *et al.* (2003); Meyer *et al.* (2012).

agents. It can be done using input–output models or general equilibrium models, although the excessive rigidity of certain models (set prices and no substitution) makes them too pessimistic, and the excessive flexibility of others (perfect market) makes them too optimistic. Approaches that steer a middle course are therefore recommended, including by the European Costs of Natural Hazards (CONHAZ) project.

The effects of macroeconomic impacts and impacts on public finances are often very different, and depend on the country's level of development, its business cycle prior to the disaster, insurance and reinsurance coverage, and other factors (von Peter *et al.*, 2012). The damage from the Christchurch earthquake in New Zealand amounted to 20% of national GDP without significantly affecting either growth in GDP or public finances (Laframboise & Loko, 2012), whereas the tsunami of March 2011 and the Fukushima crisis in Japan, where estimated damage was 3.5% of GDP, led to a 0.7% contraction in Japanese GDP and a 9.5% increase in the deficit in 2011 (OECD, 2014a, b). Macroeconomic modelling is necessary for *ex ante* evaluation of this type of impact. This requires tools and databases to relate the effects of the disaster on a national or possibly a regional scale, in cases where this type of modelling is appropriate. Analysis of public finances must also have regard to budgetary contingency mechanisms (Phaup & Kirschner, 2010).

2.5. *Assessing the overall microeconomic effects with flood scenarios*

For this purpose, a hybrid approach was developed, combining modelling of direct damages, assessment of the impacts connected with the interruption of critical networks and macroeconomic modelling with a General Equilibrium model. Three flood scenarios centred around the 100-year flood occurrence were used, taking threshold effects into account (Table 4).

Scenario S1 assumes 80% of the flood intensity of the 1910 event. The waters of the Seine reach a height of 6.90 m on the Austerlitz gauge and stay high for a week. Thanks to well-maintained local defences, the water in the major areas of urban development is largely contained by the protective banks and walls along the Seine. Nonetheless, the flood affects more than 100,000 people and floods 30,000 homes in the outer suburbs upstream and downstream of Paris. Businesses and public facilities are also directly affected, resulting in operating losses and local disruption to water treatment, electricity and transport networks for two further weeks at least. Although slightly less severe, this scenario is representative of the historic floods of January 1924 and 1955.

Table 4. Features of the flood scenarios used.

Features	S1	S2	S3
River discharge (/1910 flood)	80%	100%	115%
Maximum water height (Austerlitz)	6.90 m 1924 flood (7.32 m)	8.15 m	9.11 m 1910 flood (8.62 m)
Flood duration	1 week	2 weeks	1 month
No. inhabitants affected	100,000	600,000	1,000,000
Critical networks • Electricity • Transport • Water	Localised partial disruption for 2 weeks in Paris suburbs	Extensive disruption for 2 months in Paris suburbs	Universal disruption: • 1.5 million customers without power • Metro lines closed for several months • 6.5 million customers without water (or with substandard water)
Disruption to economic activity	2 weeks	1–2 months	2–5 months

Sources: Interviews with EPTB Seine Grands Lacs, Secrétariat Général de la Zone de Défense et de Sécurité de Paris, Direction Régionale et Interdépartementale de l'Environnement et de l'Énergie.

As a result of the works subsequently undertaken on the riverbed and bridges and the action of the reservoirs, the water in scenario S2 is 50 cm below the 1910 level, even though the river's discharge represents 100% of that of the 1910 event. The water height of 8.15 m at the Austerlitz gauge means that the City of Paris is protected by its banks and walls but Paris suburbs are not, and the water overtops the defences there and spreads over a large area. Over 600,000 inhabitants are directly affected and around 100,000 homes are flooded. The waters remain high for around two weeks, causing serious damage to businesses in the flood-prone area and a great deal of damage to public facilities and network infrastructure. However, the fact that the defences protect the core urban area means that most functionality in the metropolitan area is restored within 1 or 2 months of the end of the crisis period.

Scenario S3 envisages a 15% rise in intensity compared to the 1910 flood, and the water level reaches 9.11 m at Paris Austerlitz. As a consequence, the banks and walls are no longer capable of containing the water in several districts of Paris, and extensive areas of the urban area are submerged. The waters remain high for around 1 month in some places, directly affecting 1,000,000 people. Power cuts affect around 1.5 million households and businesses, and drinking water is no longer distributed or its quality is impaired for around 5 million inhabitants. The underground public transport network is extensively damaged and disrupted for a long period of time, and it is impossible to travel from one side of the Seine to the other. Most of the inhabitants of Île-de-France experience a severe deterioration in their living conditions for several months. This major region of Europe experiences a large disruption to its economic activity for several weeks, and it takes months to restore normal living conditions that allow the inhabitants to return to work at pre-crisis intensity. In the meantime, many small and medium enterprises (SMEs) may have gone bankrupt and large companies have (provisionally) re-located their business.

For each scenario, all direct damage and business interruptions directly linked to the flooding and to the interrupted networks were calculated first. A combined hydraulic-economic model calculated the damage for private individuals and companies using damage functions on the basis of geo-distributed data on land use and types of firms. It also assessed the business losses suffered by firms in the floodplain. The combination of the vulnerabilities and damage to various critical networks is based on the work carried out with the network operators on crisis management for a decade. The effect on firms' activity due to the interruption to electricity and transport networks was assessed using the proxy of the number of employed affected, in relation to those directly affected by the flood. By integrating all these microeconomic effects, a coherent and holistic initial economic assessment of direct and indirect damages and losses was obtained. Results were regrouped in terms of the destruction of public and private capital stock, as well as reduction of business turnover according to the various crisis scenarios (Table 5).

2.6. The macroeconomic impact of a major shock could be significant in terms of GDP, employment and public finances

With regard to the macroeconomic impact, a dynamic general equilibrium model was developed to assess the indirect effects on growth, employment and public finances, and to incorporate non-linear

Table 5. Microeconomic impact of flood scenarios (direct and indirect impacts in billion euros).

		S1	S2	S3	
		Flood: 1 week Indirect impacts: 2 weeks	Flood: 2 weeks Indirect impacts: 2 months	Flood: 1 month Indirect impacts: 2–5 months	
Destruction of capital stock	Housing and moveable property	0.76 (of which 0.04 automotive)	4.08 (of which 0.23 automotive)	6.83 (of which 0.39 automotive)	
	Stock (businesses) Equipment (businesses) Buildings (businesses)	0.81	4.71	8.54	
	Destruction of private capital stock TOTAL	**1.57**	**8.79**	**15.37**	
	Networks:				
	– Electricity	0.25	0.50	1.00	
	– Public transport (subway)	–	1.00	5.00	
	– Public transport (train)	1.00	2.00	5.00	
	– Roads	0.00	0.05	0.10	
	– Water	–	–	1.00	
	– Public facilities and buildings	0.35	1.12	1.93	
	Destruction of public capital stock TOTAL	**1.60**	**4.67**	**14.03**	
Operating losses (fall in turnover)	SME failures (annual impact)			A	B
		–	–	1.25	3
	Total losses (quarterly impact) including:	**0.58**	**5.67**	**12.33**	**16**
	– Direct effects of flooding	0.19	1.06	1.96	1.96
	– Effects of disruption to power supplies	0.19	2.59	4.67	5.4
	– Effects of disruption to transport	0.19	2.02	5.7	8.64

Note: Scenario 3B is more severe than scenario 3A, with more SMEs going bankrupt and critical network interruptions lasting a longer period.

effects; a national model enabled the impact to be represented dynamically in the short, medium and longer term. The incorporation of the specific features of the compensation funding linked to the French natural catastrophe insurance system also made it possible to assess the impact on public debt and to test various budget responses to such a catastrophe.

Assessments of the economic impacts of various flood scenarios centred around the 100-year flood of 1910 show that a large-scale shock could have a significant macroeconomic impact in terms of GDP, with repercussions both on employment and on public finances. These could come under severe pressure with corresponding deterioration over a long period. According to flood scenarios, the damage from such a catastrophe has been estimated to be between 3 and 30 billion euros for direct damage, together with a significant reduction in GDP which, over 5 years, could reach 1.5 to 58.5 billion euros, i.e. a consolidated total of 0.1–3%. The resulting contraction in business activity could have a significant effect on the demand for labour; up to 400,000 jobs could be lost in the worst case scenario. Even if a rebound in business activity could rapidly reduce some of these effects after a year, the harmful consequences of a major Seine flood could be felt over the medium to long term and weigh on public finances (Figure 3). In the case where the impact exceeds the reserves available through Cat-Nat (the national catastrophe compensation regime) and the Central Reinsurance Fund, the State could be called upon to fully assume its role of ultimate guarantor.

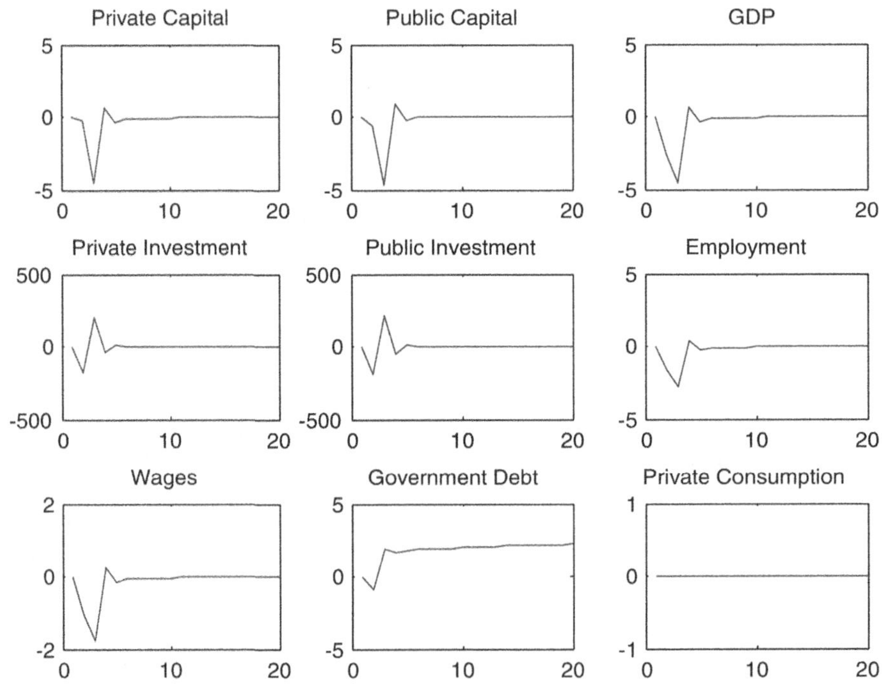

Fig. 3. Macroeconomic impact of a flood scenario over 5 years. *Note*: these charts show the variation of the different parameters as a percentage of the initial state as a function of time with quarterly measurement. The flood occurred during the first quarter. The flood scenario represented (S3) corresponds to a flood with the same water-height as the 1910 one, with a 15% greater flow.

Even if these effects are significant, it should be emphasised that this analysis is an exploratory one, and that it does not implicate a systemic risk with irreversible effects: a variety of budgetary response mechanisms could rapidly be put in place – if they are foreseen and planned in advance. Nevertheless, there is considerable uncertainty and the effects could also be accentuated by the impact of a flood on the rest of the Seine River basin.

2.7. Ambitious prevention policies could boost the resilience of Île-de-France in the long term

In light of these issues, such a situation is in no way inevitable: an effort to recalibrate, better co-ordinate, and refocus public policies would decrease the consequences of the risk whilst increasing resilience. The aim is to increase the capacity of the Île-de-France ecosystem to restore its functional capacities rapidly, in both human and economic terms. Apart from longstanding investments made over the last century, additional foresight and investment efforts may enable reduction of the risk and the vulnerabilities. Prevention policies could be strengthened along with better governance and appropriate financing mechanisms commensurate to the resources and economic advantages of this strategic territory.

3. Governance of flood risk prevention of the Seine River in the Île-de-France region

3.1. Towards a shared strategy: efficient distribution of roles and responsibilities

Response to the major risk of the flooding of the Seine in the Île-de-France region should be based on an appropriate governance framework aimed at organising public policies to improve resilience. Experience in other OECD countries shows that risk management requires the co-ordination of a large number of organisations and resources at various administrative levels, from the private sector and civil society. For each of the phases of the risk management cycle, a robust legal and institutional framework, together with well-defined governance mechanisms, is necessary to enable an integrated approach for risk management. Hence, an effective prevention policy must be based on a clear definition of the obligations of each, supported by incentive and sanction mechanisms to effectively reduce exposure and vulnerability. The key elements of good governance concern the coherence of the legal and regulatory framework and of institutions' mandates to contribute to an established and shared strategy, as well as co-ordination and effective co-operation between the various stakeholders for its implementation. This includes questions of vertical coherence – between the various administrative levels – and horizontal coherence – between the various spheres of public policy – in the distribution of roles and responsibilities, avoiding duplications of effort and favouring synergies (Charbit, 2011; OECD, 2012).

3.2. The institutional context has not favoured the emergence of an ambitious and coherent strategy for preventing Seine River floods in Île-de-France

Despite a progressive body of legislation (Laws of 1982, 1995 and 2003) and an exemplary set of regulatory, financial and contractual tools (Risk Prevention Plans, Cat-Nat insurance regime, Prevention Funds, Flood Prevention Action Programmes) for the prevention of risks at the national level (Asca, 2012), the Seine flood risk in the Île-de-France region is managed within a fragmented framework resulting from successive waves of decentralisation. Despite the risks faced by the region, this has partially hampered the emergence of a global shared vision on flood risk management. The lack of an overall vision in this strategic territory – contrary to other major French rivers such as the Loire (OECD, 2010) or the Rhône – reveals a governance weakness, even if awareness of the issues at stake is emerging. The policy tools developed at a national level have in the past struggled to find a practical and effective application in this region with extraordinary issues at stake.

3.3. Institutional fragmentation has, in the past, been a constraint on action

With the prospect of establishing an holistic strategic framework for managing the risk of floods of the Seine, a particularly acute question of governance is raised in Île-de-France. The institutional and territorial fragmentation in flood prevention results in poor governance particularly among the various administrative levels. As a rule in France, mayors and prefects are solely responsible for managing risks, both with regard to prevention and crisis management. In the case of the Île-de-France region, its specific nature and resulting institutional characteristics add another layer of complexity to decision making. The large number of stakeholders involved (whether national, regional, departmental, municipal or metropolitan) makes it difficult to ensure synergies between the various administrative levels. Apart from questions of multi-level co-ordination, territorial disparities prevent the emergence of a shared vision. Competing views may in fact appear between Paris and its suburbs, the West and the East of the territory at risk, the urban area and the peri-urban and rural areas. Each of these areas has different levels of exposure. Their technical, financial and human capacities to implement public policies independently also vary.

3.4. Overcoming administrative fragmentation to facilitate interaction between various public policies

Beyond questions of horizontal and vertical co-ordination between the various administrative levels, there is also a need to address several areas of public policies which contribute to the various dimensions of flood risk prevention. Each of these policies involves specific stakeholders, different territorial sensibilities, as well as different approaches which may confront or be in conflict with each other. Up to now, isolated approaches have often prevailed. The various policies include:

- the risk prevention policy conducted by the Ministry of the Ecology, Sustainable Development and Energy (MEDDE) and its decentralised structures in the Île-de-France region;
- the crisis management policy of the Ministry of the Interior (and its Prefecture de Police – Paris Defence area), with a large number of stakeholders in the Île-de-France region;
- the territorial development and planning policy in which the regional council plays a key role as well as governance at the local level through town planning;
- the water management policy and its institutions, including the Seine-Normandy Water Agency (Agence de l'Eau Seine-Normandie), which plans and finances the conservation of water resources in the Seine basin, and EPTB Seine Grands Lacs, an operational stakeholder which historically manages four large dams upstream of the Seine basin with the double objective of combating floods and supporting low water levels.

3.5. Well-identified governance deficiencies can be overcome

The governance of flood risk management and prevention in Île-de-France appears to be very complex. Because of their dispersion, existing efforts cannot be fully effective. The inadequate distribution of responsibilities and resources among stakeholders at various levels has prevented the emergence of a coherent leadership as well as a common vision with shared objectives for the prevention of flood risk. Strategic planning documents on the Seine basin, on the river development, or on the development of the Île-de-France region have not up to now allowed for a genuine multi-stakeholder approach or to align the various initiatives in a common strategy for flood risk prevention. Only the work on crisis management has been able to unify metropolitan area stakeholders on the development of an emergency response plan.

An appropriate linkage between the two spatial scales – that of the river basin for work on the hazard side and of the exposed metropolitan area for work on the vulnerability reduction side – is a condition for the successful implementation of effective prevention policies. Furthermore, despite the involvement of a large number of stakeholders, there is no criterion that makes it possible to assess the respective contributions of the preventive measures undertaken by each stakeholder. This lack of performance assessment increases the difficulty in efficiently allocating responsibilities and resources for risk prevention. Subsidiarity, local ownership, monitoring and evaluation of the measures undertaken, as well public engagement and participation in decision-making, are the principles of good governance which would enable increased responsibility and accountability of the various stakeholders.

3.6. Opportunities particularly from the Greater Paris project and the EU Floods Directive, should be seized

Today, there is on-going momentum in flood risk management with the implementation of the European Directive on assessing and managing flood risks, for which 2013–2015 is a key period: a national strategy on the management of flood risks has recently been developed and a priority area for flood risk management in the Île-de-France region was recently defined (MEDDE, 2014). Comprised of 141 municipalities, this High Risk Territory (*Territoires à risque important d'inondation* – TRI) seems to be the appropriate scale to address issues of vulnerability. A flood risk management strategy must be developed by 2015 in the TRI, together with a governance mechanism for its implementation. This is being established, under the auspices of the State, through a partnership between both prevention and crisis management actors. Local stakeholders such as EPTB Seine Grands Lacs will be associated to this partnership within its area of jurisdiction. Furthermore, a Flood Prevention Action Programme (*Programme d'Action pour la Prévention des Inondations* – PAPI) was also prepared by EPTB Seine Grands Lacs and local stakeholders, which anticipates substantial work on the hazard control side and additional actions on the vulnerability of the Île-de-France urban area.

Opportunities are also arising to incorporate resilience into the Greater Paris development project. This long-term investment project for the whole metropolitan area makes it possible to tackle the issue of flood risk at the scale of the Île-de-France urban area and to take major urban projects into account. These opportunities could make it possible to engage the region in an ambitious and long-term resilience approach largely shared with all the stakeholders. Opportunities for the development of the Greater Paris development project in the coming decades can be fully met through a transparent and responsible approach to risk management questions.

4. Increasing the resilience of Île-de-France by flood risk prevention

The only way of reducing the Seine flood risk in the Île-de-France region is by means of practical measures aimed at increasing the territory's resilience. Even though re-examining governance will allow a vision, objectives and major principles of a flood management strategy to be determined, their implementation should take shape at the local level (in the upstream basin, in exposed territories, in public and private planning and development projects, within companies) for two major areas of action: hazard control and vulnerability reduction. If structural measures aimed at limiting risk exposure by building infrastructure were given priority in the past, their financial, social and environmental limits are now leading towards risk control approaches that are more aligned with environmental protection.

Reduced vulnerability can also be achieved through non-structural measures. It is fundamental that risk knowledge and awareness be developed to create a culture of risk as a condition for action at any level. Increasing territories' resilience may be based on a more balanced urban development which incorporates flood risk appropriately. This includes the question of the critical networks and infrastructures whose vulnerability to floods creates the potential for cascading knock-on effects. On a wider scale, the resilience of firms and individuals should also be developed, for instance through approaches aimed at business continuity. Finding the right mix between structural and non-structural prevention measures is the challenge explored in this section, as per the OECD Recommendation on the Governance of Critical Risks (OECD, 2014a, b).

4.1. Risk awareness is progressing thanks to harmonising approaches

Flood risk awareness is growing in Île-de-France and the on-going harmonisation of approaches will in time allow all actors involved in risk prevention to have the information required to enable them to act coherently. Up to now, the multiplicity of approaches, tools and standards for assessing risks have played their part in causing some confusion, preventing stakeholders from agreeing on similar results, with each of them tending to develop their own evaluation methods. The current process of sharing and harmonising awareness among actors in risk prevention and crisis management – including network operators – as well as developing an accurate mapping of the risks with the implementation of the European Floods Directive, makes it possible to envisage having the tools required to devise and make a detailed assessment of all the preventive measures. This could be continued with other stakeholders, such as those in the insurance sector, in a consistent comprehensive risk assessment approach, particularly from an economic standpoint. Initiatives carried out at the national level can contribute to improving this situation, particularly by the creation of the National Observatory of Natural Risks together with the insurance sector.

4.2. Reinforcing the risk culture among citizens and decision-makers is essential as the memory of historical floods fades

Citizens' and decision-makers' risk perception is very low. According to all stakeholders, the level of information and degree of awareness of a major flood risk amongst citizens are insufficient, in view of the risk level. There has not been a significant flood for almost 60 years and the impacts of low frequency events of 10 to 30 years ago have nearly disappeared thanks to the upstream dam reservoirs. As a consequence, flood risk tends to fade from the collective memory. On the other hand, the collective insurance coverage provided by the French Cat-Nat compensation scheme, while presenting many advantages, can create a moral hazard by giving citizens, firms and decision-makers the impression that, come what may, they will be compensated for their damage: this does not lead to increased risk awareness or to initiatives to try to set up preventive measures. A voluntary approach to raising awareness in Île-de-France is necessary to develop a risk culture.

4.3. Effectiveness of risk communication is limited while private sector awareness slowly increases

The development of a flood risk culture seems to be a priority for many stakeholders at the local level but the effectiveness of risk communication appears limited. French risk regulations make procedures to inform citizens about the risk they are exposed too. However, they do not institute quantified objectives for risk awareness. At the same time, non-regulatory innovative approaches for raising citizens' awareness have met with a good response from the population (exhibitions, work in schools, 3D films). Overall, it would appear that the public decision-makers' failure to communicate on the subject is a major factor limiting the development of a culture of risk. This reveals a low awareness of the risk, since flooding continues to be considered unlikely.

In companies, awareness has increased in recent years but is still variable. Large companies in a certain number of sectors (energy, transport, water, telecommunications, banking) have become aware of this risk through their participation in crisis preparedness or through the regulation on business continuity for vital sectors. Conversely, smaller companies or other sectors have only a very limited awareness of the risk. There are very few actions intended for small businesses, nor toward essential stakeholders for resilience development such as those in engineering, urban planning or architecture. Ultimately, the differences in the degree of awareness and commitment between the various sectors and the various stakeholders hamper the development of a genuinely shared culture of safety.

4.4. Control of urban development in the floodplain comes up against limited regulatory tools

With regard to territorial resilience, risk prevention policies based on controlling urban development have only limited effects in respect of the underlying economic issues at stake in the Paris area. The sharing of responsibilities for land use planning prevents a coherent management of the risk. The local development approach, instead of encouraging municipalities to limit building in the floodplain, encourages them to develop these often very attractive areas. In this context, where the fabric of the urban areas does not put the flood issue at the forefront of its planning process and does not set any objectives, public decision-makers and planners are ultimately content to live with the regulatory aspects, particularly through the Flood Risk Prevention Plans. These documents negotiated by the State and the local authorities only determine a few non-constructible zones; they are not standardised across the different areas at risk in the basin, do not impose specific regulations to network operators and are not restrictive with regard to existing constructions, which are predominant in Île-de-France. In this way, over the past 20 years, 1,500 hectares of the floodplain have been built on, including some major infrastructures (IAU, 2011a, b).

4.5. Urban planning and innovation policies could use the Greater Paris project as an opportunity to boost resilience

The Greater Paris project includes urban densification and the development of a major transport network by 2030. In the context of a densely populated urban area, this project offers opportunities: a flood-resilient city may emerge from innovative urban projects built along the Seine. Examples in other OECD countries show that resilience can be the source of innovation and, in this way, participate in green growth (OECD, 2013). Certain initiatives have proved that it is possible to build an urban environment with infrastructures that are resilient to floods or to improve existing constructions when a strong political will and a sustainable governance structure carry this ambition. Resilience to floods is at the heart of the project for the urban renewal of the Ardoines district, severely exposed to this risk. Located in the area of the Orly-Rungis-Seine-Amont Development Agency, this project is directly managed by the State and may serve as an ambitious demonstration of resilience innovation (Brun & Adisson, 2011). Similarly, initiatives around green and blue corridors in the Île-de-France region may incorporate flood prevention.

4.6. Investment in improving the resilience of critical networks and infrastructures will be key to ensuring resilience for the whole metropolitan area

Urban networks and critical infrastructures structure the region and enable it to function on a daily basis (electricity, water, telecommunications and transport). Despite the existing regulatory or contractual context, it appears that the requirements for business continuity are not sufficiently high to allow rapid recovery in the event of a major flood. Work on crisis management has contributed to raising the operators' awareness, to persuade them to assess their vulnerability, and the cascading effects they could create for other business sectors. Nevertheless, there are still great differences in their awareness levels, preparedness and resilience. Some have a precise flood risk assessment, have developed Business Continuity Plans and have sometimes invested significantly to reduce their vulnerability (including relocation). Others, however, have made few efforts or are even reluctant to share their information which could be useful for everyone's resilience. There is no precise standard or overall harmonisation which would determine the resilience and protection levels required and which would measure them with predetermined indicators. Investment in infrastructures planned for the next 30 years could be used to improve the networks' resilience.

4.7. Efforts made to increase resilience and improve public services and business continuity seem limited

With regard to companies and public services, the development of business continuity and investment in prevention is in its early stages. To ensure the continuity of government, plans must be developed by all ministries. Furthermore, resilience processes in local authorities and particularly the municipalities are limited: less than 40% of the municipalities concerned have developed a continuity plan. If a major flood were to happen, it is uncertain whether a large number of public services would continue to function. The private sector, and in particular large companies, is increasingly encouraged by the markets to take into account its risk exposure, the possible effect on the business plan and the measures likely to reduce the risk. The commitment made by companies in the private sector to improve their own resilience seems to be related to their size or to their sector. While some large companies have already developed or are currently developing their own prevention and flood risk strategies in accordance with the regulatory framework and the regulation authorities (banks, telecommunications), overall, SMEs are still vulnerable and ill-prepared.

4.8. Levels of protection against floods are not harmonised at the Paris urban area scale

The difference between the levels of protection provided by dykes and clay walls, and the relative levels of maintenance and investment between the centre and the outskirts of the urban area do not ensure uniform protection for the citizens of Île-de-France, reflecting the historic layers which no longer correspond to today's urban and industrial density (Figure 4). Recent efforts to analyse their vulnerability under the auspices of the State should be emphasised, as well as the reinforcement work carried out in some places, albeit in a somewhat fragmented approach due to separate contracting authorities. Contrary to other OECD countries, the lack of any predefined standard level of protection accentuates the negative effects of the lack of any overall management approach for these protections. It does not enable the level of risk investment required to be determined at the basin level.

4.9. Hazard control depends on effective management of upstream multiple-use reservoir-lakes

The flood hazard is also managed by four reservoir-lakes built upstream of the basin. With a storage capacity of 800 million m^3, these big dams can together lower the water level by 70 cm in Paris and reduce direct damage by half (EPTB Seine Grands Lacs, 2013). They are run by EPTB Seine Grands Lacs whose mission is gradually being extended to other aspects of integrated water resources

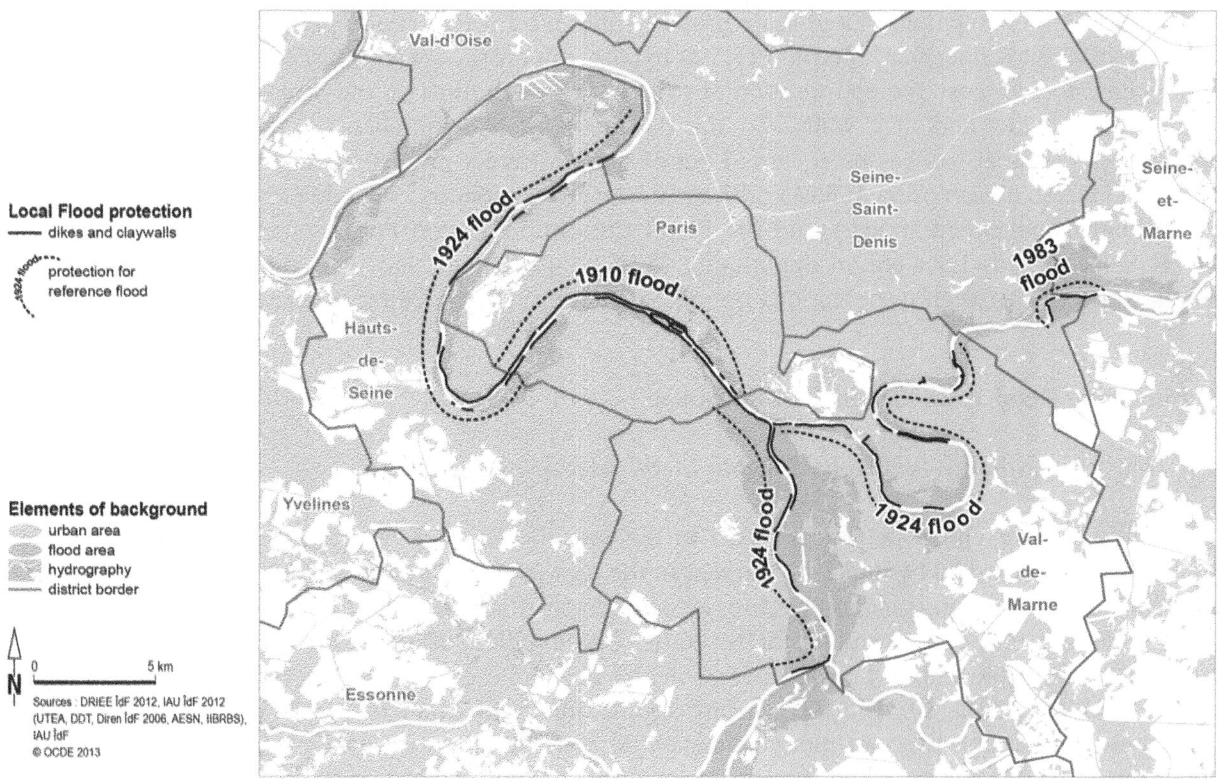

Fig. 4. Location of dykes and clay walls in Île-de-France.

management. Since their construction, in the absence of major floods, the functions of these reservoir-lakes has become somewhat focussed on these other uses (low-flow support, leisure activities). The establishment of a new fee collected for the low-flow support service to the major water users will contribute to the EPTB budget, strengthening this part of its mandate. The optimisation of the management of existing reservoirs with respect to different uses represents a key issue which should be regularly discussed, particularly in the context of climate change.

4.10. The issues of an innovative but costly new infrastructure in La Bassée

A new hazard reduction project has emerged, the La Bassée project, which raises questions on the funding of and priorities in hazard control actions and governance. The innovative approach of this project consists of pumping water from the Seine, before it is joined by the tumultuous Yonne River, into storage basins installed along the river. The project was developed with a state-of-the-art approach, including the consultation of local populations upstream through a large and transparent public debate, the incorporation of multiple uses in its design (restoration of wetlands, eco-tourism, economic activities), a positive cost–benefit analysis and a multi-criteria analysis. However, in order to better justify its cost-effectiveness, this project must still demonstrate its operational utility and respond to some key questions related to the definition of its operational rules and related decision-making in times of crisis. The idea to realise this project in stages seems interesting insofar as each stage will enable experimentation to test the structure to reduce the water level in the event of a major flood.

Apart from this potentially large-scale project, other hazard reduction options have also been identified, such as renovating the Joinville-Le-Pont bypass valve which would protect a large number of the inhabitants at a low cost, the optimisation of existing infrastructures, or the ecological restoration of the basin heads, dykes and clay walls. To the detriment of a comprehensive and genuinely efficient approach, no comparison has been carried out of these different projects in terms of costs and benefits due to the fact that each of them has a different contracting authority. Together, the establishment of the local flood management strategy and the Flood Prevention Action Programme project supported by EPTB Seine Grands Lacs are an opportunity to make a reasoned and transparent choice *vis-à-vis* all the stakeholders between these various options, including non-structural measures.

5. Financing flood resilience in Île-de-France

5.1. How can resilience be financed within a constrained budget context?

Funding the preventive actions required to increase the level of resilience remains a major issue. Within a context marked by under-investment in the past, and the current difficult economic conditions, investment in prevention is under pressure, given concerns for balancing the budget and the necessity to decide priorities for public spending. In the Île-de-France region, as often elsewhere, decisions to undertake and to fund prevention are dependent on the economic context or the triggering role of recent events (OECD, 2014a, b). The lack of any significant flood event over the past 60 years tends to reduce awareness and does not motivate stakeholders to structure a financial approach to prevention challenges. The differences in the levels of risk and the amount of prevention efforts between the different exposed areas of Île-de-France make it difficult to fund infrastructures which would be of greater benefit to some rather than others. This leads to a lack of action for funding more collective resilience. The allocation of prevention resources is a challenge which requires demonstration that public funds are used as efficiently as possible.

5.2. Tools for financing flood risk prevention exist, particularly through insurance

For 30 years now, France has engaged in significant flood risk prevention efforts. A series of innovative tools has been set up with related funding mechanisms. Funding flood prevention is generally based on solidarity mechanisms. In addition to national budget resources, France has established an original collective insurance mechanism, the Cat-Nat insurance regime, based on a public–private partnership between insurance companies and the State and on the solidarity principle – among all the insured – against the risks of natural catastrophes. This system also makes it possible to make a large contribution to risk prevention funding, particularly for flood risks, without a direct impact on public finances. Flooding is the most frequent as well as the most damaging natural disaster in France. Nevertheless, this system is faced with growing demands.

Additional resources come from local authorities, which are generally mobilised through contractual tools with the State such as the Large River Plans and the Flood Prevention Action Programmes (PAPI) or to fund basin organisations such as EPTB Seine Grands Lacs which manages the Seine's upstream reservoir-lakes. Other sources of finance can also be used for prevention, such as those linked to wider water policy like the water agencies, network operators and companies (which could invest in their own resilience), or the European Union (particularly to implement the EU Floods Directive).

5.3. The financing of flood prevention in Île-de-France has benefited from limited existing resources

With annual average damage from the Seine flood risk equivalent to one quarter to one third of annual average damage caused by floods in France – 1 to 1.4 billion euros (MEDDE, 2012) – it would seem relevant to match the prevention efforts to this level of risk. Expenditures on flood prevention in France may be assessed at between 300 and 450 million euros, or about one third of the estimated damage (MEDDE, 2013). Such a level of investment in prevention may be considered satisfactory in view of the public expenditure effectiveness criteria, if it is ensured that these funds are allocated as a priority towards the preventive measures which have the greatest benefits. Although important investments allowed the upstream reservoir dams to be built in the past century, it appears that the instruments for funding prevention have played a very small part in reducing the vulnerability of Île-de-France over the last 20 years, compared with other regions or river basins. Other strategic priorities, in particular measures to protect human lives, have mobilised authorities and the available prevention funds. This has led to a certain delay in funding the prevention of this major flood risk, as a flood of the Seine would have limited casualties. Up to now, the national allocation of resources has not been based on criteria giving priority to resources according to the level of risk. This is changing with the implementation of the EU Floods Directive which identified the most risk-prone territories, of which the Paris metropolitan area is one.

5.4. The development of a financing strategy may be based on principles aligned with improved governance

There is room for progress in defining a financing approach for prevention which is adapted to the issues at hand. In the context of strained public finances, the question of additional resources and the sharing of efforts (State, local government, businesses, citizens, European funds) may be addressed by setting out a number of principles for an overall financing strategy. Improved governance of risk management could help define a level of risk as a shared objective upon which a financial approach could be developed. The general principle is that the beneficiaries of prevention measures should be the first to finance prevention (OECD, 2003). Identification of the beneficiaries helps to determine the primary sources of funding to be raised for such a strategy. Prevention funding must aim at being most effective through an economic approach based on:

- *Coherence:* heterogeneity of the different stakeholders' preventive approaches results in duplicated actions and extra costs, as well as uncooperative 'free-rider' behaviour and distortions of competition and levels of service.
- *Economic efficiency:* a generalisation of cost–benefit studies and multi-criteria analysis, apart from judging the relevance of a specific project, could allow a comparison to be made of the various options available and their benefits, including non-structural measures.
- *Long term:* long-term investment planning makes it possible to introduce flexibility in choices, to take into account the evolution of knowledge and to reduce uncertainties so as to adjust resources in accordance with needs.
- *Equity:* this question arises both regarding the strategic allocation of national resources for this territory exposed to significant risk, as well as allocation within the Île-de-France region, in view of the differences between the levels of protection.

5.5. Existing and additional resources

A large number of existing financial mechanisms may be further mobilised to prevent this major risk. Adopting a multi-hazard approach (flood, drought, pandemic, terrorism) can provide access to the funding of a water policy or for the management of risks in the wider sense. A long-term approach in relation to the Greater Paris regional development project also opens up avenues. Many European Union mechanisms also provide funds for risk prevention and this ought to be explored. Several potential sources of additional finance could be mobilised as well. A certain number of actors in the private sector would be prepared to provide resources if they are shown that their contribution to investing in prevention could significantly reduce their level of exposure to the risk. Existing taxes on added-value real estate in the flood zone, local taxes or those on the tourist sector could also be explored as sources of funds. New resources in the form of service fees established for low-flow support by EPTB Seine Grands Lacs could also inspire a similar system benefiting flood protection, particularly for network operators.

6. Policy recommendations

This review of public policies for the prevention of the Seine flood risk in the Paris Île-de-France region has led to the development of policy recommendations that the French authorities may wish to consider to set up an ambitious resilience strategy for this major risk, as listed below. An inclusive discussion with all the key stakeholders as well as a peer review with other OECD countries in the framework of the OECD High Level Risk Forum has ensured their relevance to the challenges identified in this analysis.

1. Ensure the appropriate linkages between the various levels of flood prevention – from the exposed Île-de-France metropolis to the river basin.
2. Define an ambitious and mobilising global vision over the long term together with actionable principles.
3. Break down the global vision into precise objectives and make the stakeholders aware of their responsibilities.
4. Create effective gateways between the flood risk management strategy and related public policies.
5. Continue to improve and harmonise risk knowledge and ensure that risk information is made available.
6. Reinforce the risk culture of citizens, decision-makers and companies.
7. Improve territorial resilience, using the opportunities offered by the Grand Paris project.
8. Gradually improve the resilience level of critical networks and take steps towards preserving the continuity of business and public services.
9. Place the flood protection infrastructures under the responsibility of a single contracting authority.
10. Encourage experimentation with regard to the La Bassée storage project.
11. Support the local Seine flood risk management strategy in the Île-de-France with a clear financial strategy.
12. Mobilise all the beneficiaries of preventive measures in a multi-level approach.
13. Strengthen efforts to clarify the priority criteria for prevention funding from State resources.
14. Re-examine the impact of the Cat-Nat compensation scheme on flood risk prevention.

Disclaimer

The opinions expressed and the arguments employed in this paper are those of the author and do not necessarily reflect the official views of the OECD or of the governments of its member countries.

References

Aon Benfield (2012). *2011 Thailand Floods Event Recap Report*. Aon Benfield (Thailand) Ltd, Bangkok.

AScA, Ledoux Consultants (2012). *L'agence de l'eau Seine-Normandie et la gestion du risque inondation: quelle stratégie de positionnement? Synthèse stratégique*. Agence de l'eau Seine Normandie, Nanterre.

Brun, A. & Adisson, F. (2011). Renouvellement urbain et risque d'inondation: le plan guide Seine Ardoines. *Cybergeo: European Journal of Geography (online)*. Aménagement, Urbanisme, document 561, 29 October 2011. http://cybergeo.revues.org/24751.

Caisse Centrale de Réassurance (2013). La crue de la Seine en Île-de-France, Étude historique de la crue de 1910 - Modélisation de scénarios de référence. (Internal document). Caisse Centrale de Réassurance, Paris.

Chambre de Commerce et d'Industrie Paris Île-de-France (2013). Chiffres-clés de la région Île-de-France 2013. Paris. www.crocis.cci-paris-idf.fr/uploads/_crocis/ani_fichiers/crci_france_2013_hd.pdf.

Charbit, C. (2011). Governance of Public Policies in Decentralised Contexts: The Multi-Level Approach. *OECD Regional Development Working Papers*, 2011/04, OCDE, doi: http://dx.doi.org/10.1787/5kg883pkxkhc-en.

Cooper, C. & Matsuda, K. (2013). Tokyo prepares for a once-in-200-year flood to top Sandy. *Bloomberg*, 31 mai 2013.

Ducharne, A., Sauquet, E., Habets, F., Deque, M., Gascoin, S., Hachour, A., Martin, E., Oudin, L., Page, C., Terray, L., Thiery, D. & Viennot, P. (2011). Évolution potentielle du régime des crues de la seine sous changement climatique. *La Houille Blanche 1*, 51–57, http://hal.archives-ouvertes.fr/docs/00/59/37/80/PDF/LY2011-PUB00031416.pdf

Environment Agency (2012). *Thames Estuary 2100: Managing Flood Risk through London and the Thames Estuary*. Environment Agency, London.

EPTB Seine Grands Lacs (2013). *Programme d'Actions de Prévention des Inondations de la Seine et de la Marne Franciliennes*. Rapport de présentation, EPTB Seine Grands Lacs, Paris.

IAU (Institut d'Aménagement et d'Urbanisme) (2011a). Économie francilienne: quelle robustesse face à une inondation majeure? *Note rapide économie, no. 534*. IAU, Paris, www.iau-idf.fr/fileadmin/Etudes/etude_784/NR_534_web_01.pdf.

IAU (Institut d'Aménagement et d'Urbanisme) (2011b). Urbanisation et zones inondables: les risques encourus. *Note rapide territoires, no. 557*. IAU, Paris, www.iau-idf.fr/fileadmin/Etudes/etude_839/NR_557_web.pdf.

Japan International Cooperation Agency (JICA) (2012). JICA's support to 'Toward safe and resilient Thailand' through revising the comprehensive flood management flood management plan for the Chao Phraya River Basin. Forum of 20 February 2012, www.unescap.org/idd/events/2012-South-East%20Asia%20Flood%20Risk%20Reduction%20Forum/JICA.pdf.

Kryžanowski, A., Brilly, M., Rusjan, S. & Schnabl, S. (2013). Structural flood-protection measures referring to several European case studies. *Natural Hazards and Earth System Sciences*, (1), www.nat-hazards-earth-syst-sci-discuss.net/1/247/2013/nhessd-1-247-2013.html.

Laframboise, N. & Loko, B. (2012). Natural Disasters: Mitigating Impact, Managing Risks. *IMF Working Paper*, no. 245, www.imf.org/external/pubs/ft/wp/2012/wp12245.pdf.

Land of Hesse (2007). *Hesse Flood Action Plan. (Landesaktionsplan Hochwasserschutz Hessen)*. Hessisches Ministerium für Umwelt, ländlichen Raum und Verbraucherschutz, Hesse.

Lin, N., Emanuel, K., Oppenheimer, M. & Vanmarcke, E. (2012). Physically-based assessment of hurricane surge threat under climate change. *Nature Climate Change*, pp. 462–467, www.nature.com/nclimate/journal/v2/n6/full/nclimate1389.html.

Meyer, V., Becker, N., Markantonis, V. & Schwarze, R. (2012). *Costs of Natural Hazards: A Synthesis – CONHAZ Costs of Natural Hazards*. UFZ, Germany.

MEDDE (Ministère de l'Écologie, du Développement durable et de l'Énergie) (2011). *L'évaluation préliminaire des risques d'inondation 2011: Bassin Seine-Normandie*. Ministère de l'Écologie, du Développement durable, des Transports et du Logement, Paris.

MEDDE (Ministère de lÉcologie, du Développement durable et de l'Énergie) (2012). *Mieux savoir pour mieux agir: principaux enseignements de la première évaluation des risques d´inondation sur le territoire français 2011*. Ministère de lÉcologie, du Développement durable et de lÉnergie, Paris. http://catalogue.prim.net/190_mieux-savoir-pour-mieux-agir-principaux-enseignements-de-la-premiere-evaluation-des-risques-d-inondation-sur-le-territoire-francais-epri-2011.html.

MEDDE (Ministère de l'Écologie, du Développement durable et de l'Énergie) (2013). Les dépenses publiques et les bénéfices de la prévention des risques naturels. Études et Documents n° 94, commissariat général au Développement durable.

MEDDE (Ministère de l'Écologie, du Développement durable et de l'Énergie) (2014). Stratégie Nationale de Gestion des Risques d'Inondation. http://www.developpement-durable.gouv.fr/IMG/pdf/140509_SNGRIetAnnexes_approuvee_BAT_cle0459ad.pdf.

Netherlands Ministry of Infrastructure and the Environment (2012). Le cadre législatif et organisationnel de la gestion du risque d'inondation aux Pays-Bas, audition au Sénat (29 May 2012). www.preventionweb.net/files/29781_lecadrelegislatifetorganisationneld.pdf.

New York City Special Initiative for Rebuilding and Resiliency (2013). *A Stronger, More Resilient New York*. City of New York, New York.

Norwegian Directorate of Water and Energy Resources (2009). Flood inundation maps. www.nve.no/en/floods-and-landslides/flood-inundation-maps.

OECD (Organisation for Economic Co-operation and Development) (2003). *Emerging Risks in the XXI Century: A Call for Action*. OECD publishing, Paris. doi: http://dx.doi.org/10.1787/9789264101234-fr.

OECD (Organisation for Economic Co-operation and Development) (2010). *Étude de l'OCDE sur la gestion des risques d'inondation: Bassin de la Loire, France 2010*. Éditions de l'OCDE, Paris. doi: http://dx.doi.org/10.1787/9789264056817-en.

OECD (Organisation for Economic Co-operation and Development) (2011). *Future Global Shocks, Improving Risk Governance*. OECD Publishing, Paris. doi: http://dx.doi.org/10.1787/9789264114586-en.

OECD (Organisation for Economic Co-operation and Development) (2012). *Meeting the Water Reform Challenge, OECD Studies on Water, (Répondre à l'enjeu de la réforme de l'eau, Études de l'OCDE sur l'eau).* OECD Publishing, Paris. doi: http://dx.doi.org/10.1787/9789264170001-en.

OECD (Organisation for Economic Co-operation and Development) (2013). *Green Growth and Cities.* OECD Publishing, Paris. doi: 10.1787/9789264195325-en.

OECD (Organisation for Economic Co-operation and Development) (2014a). *A Boost to Resilience: Governing Effective Prevention and Mitigation of Disruptive Shocks.* OECD Publishing, Paris.

OECD (Organisation for Economic Co-operation and Development) (2014b). *Recommendation of the Council on the Governance of Critical Risks.* OECD Publishing, Paris. http://www.oecd.org/gov/risk/recommendation-on-governance-of-critical-risks.htm.

Penning-Rowsell, E., Johnson, C., Tunstall, S., Morris, J., Chatterton, J., Coker, A. & Green, C. (2003). *The Benefits of Flood and Coastal Defence: Techniques and Data for 2003*, Flood Hazard Research Centre, Middlesex University.

Phaup, M. & Kirschner, C. (2010). Budgeting for disasters: focusing on the good times. *OECD (Organisation for Economic Co-operation and Development) Journal on Budgeting*, OECD, www.oecd.org/gov/budgeting/48168599.pdf.

Przyluski, V. & Hallegatte, S. (2011). Indirect costs of natural hazards. CONHAZ Costs of Natural Hazards, SMASH-CIRED. www.conhaz.org/CONHAZ%20REPORT%20WP02_2_FINAL.pdf.

Queensland Flood Commission Inquiry (2012). *Final Report.* Queensland Flood Commission of Inquiry, Brisbane.

Rose, A. (2009). A framework for analyzing the total economic impacts of terrorist attacks and natural disasters. *Journal of Homeland Security and Emergency Management 6*, (1), http://research.create.usc.edu/published_papers/78/.

Rose, A., Sue Wing, I., Wei, D. & Avetisyan, M. (2012). *Total regional economic losses from water supply disruptions to the Los Angeles county economy.* Final Report to Los Angeles County Economic Development Corporation. www.laedc.org/reports/WaterSupplyDisruptionStudy_November2012.pdf.

Severn Trent Water (2010). *Annex to June Return Board Overview.* Severn Trent Water, Coventry.

Smith, K. & Ward, R. (1998). *Floods: Physical Processes and Human Impacts.* John Wiley and Sons, Chichester.

Von Peter, G., von Dahlen, S. & Saxena, S. C. (2012). Unmitigated Disasters? New Evidence on the Macroeconomic Cost of Natural Catastrophes. *Working Paper No. 394*, Bank for International Settlements. www.ssti.us/wp/wp-content/uploads/2013/07/Unmitigated-Disasters-short.pdf.

World Bank (2012b). *Cities and Flooding: A Guide to Integrated Urban Flood Risk Management for the 21st Century.* World Bank, Washington.

Zizkova, E. (2004). Les inondations catastrophiques d'août 2002 en République tchèque. *La Météorologie*, no. 46, August 2004. http://documents.irevues.inist.fr/bitstream/handle/2042/36058/meteo_2004_46_34.pdf?sequence=1.

Integrated policies and practices for flood and drought risk management

A. Grobicki[a,*], F. MacLeod[a] and F. Pischke[b]

[a]*Global Water Partnership Organisation (GWPO), Linnégatan 87D; PO Box 24177, 104 51 Stockholm, Sweden*
**Corresponding author. E-mail: ania.grobicki@gwp.org*
[b]*Global Water Partnership Organisation (GWPO), 7 bis, avenue de la Paix, Case postale 2300, CH-1211 Genève 2, Switzerland*

Abstract

The paper argues for an integrated approach to the management of water-related disasters that becomes a full part of the political decision-making process at the earliest possible moment and focuses on preparedness, mitigating their negative impacts and also considering their positive impacts, particularly those of floods. By doing this, there is an opportunity to consider the three pillars of sustainable development, and understand the options that exist and the trade-offs that may need to be made between economic efficiency, environmental sustainability and social equity. Within the post-2015 agenda, water-related disasters are addressed by targets under a number of different Sustainable Development Goals (SDGs). As climate change becomes an ever-more-present reality, whose impacts are often experienced through water-related disasters such as floods and droughts, there is an urgent need to build disaster-resilient societies through more integrated policies and practices, including stakeholders' perspectives and a partnership approach. The paper provides stakeholder perspectives and approaches from around the world that are putting these ideas into practice.

Keywords: Disaster; Drought; Flood; Post-2015 agenda; Resilience; Sustainable development

1. Introduction

Building social resilience to water-related disasters is a slow process of evolution in thinking, negotiating and developing the frameworks to understand and manage risk. Step by step, policies and practices are being put in place and tested. At the World Summit for Sustainable Development in Johannesburg, 2002, countries agreed to develop and implement integrated water resources' management (IWRM), based on the three pillars of sustainable development: namely, social equity, economic efficiency, and environmental sustainability (UN, 2002). During the first World Conference on Disaster Reduction in 2005, the Hyogo Framework for Action 2005–2015 was agreed within the International

doi: 10.2166/wp.2015.009

Strategy for Disaster Reduction (ISDR), to make the world safer from natural hazards and to reduce disaster losses (UN ISDR, 2007). In 2007, the Conference of the Parties to the UN Convention to Combat Desertification (UNCCD) adopted a 10-year strategic plan to reverse and prevent desertification and to mitigate the effects of drought, in order to support poverty reduction and environmental sustainability (UNCCD, 2007). The 3rd World Climate Conference in 2009 established the Global Framework for Climate Services, providing climate-related information to help countries manage the risks of climate variability and change (World Meteorological Organization (WMO), 2009). In 2015 the 3rd World Conference on Disaster Risk Reduction in Sendai will set the post-2015 agenda for disaster risk reduction (DRR), linking DRR with the post-2015 development agenda and the Sustainable Development Goals (SDGs) soon to be agreed through to 2030. These global milestones all reinforce the interconnections between water-related disasters such as floods and droughts, integrated approaches to water policy and water management, and long-term sustainable development.

Of all water-related disasters floods and droughts stand out as two sides of the same coin, yet they are much more complex than simply a question of too much or too little water. Both have deeply rooted social and economic dimensions that tend to dominate our thinking, and dominate the way that we think of water resources' management. There is evidence that the level of river basin development itself changes the social and political perception and debate about these issues, increasing the economic emphasis. One consequence of this is that societies lose sight of the inherent value that flood and drought cycles can have in an ecological sense, and in the context of long-term river basin development. Day to day, decision-makers are facing the hard reality that managing the uncertainties of an increasingly variable climate is vital to socio-economic development. Investing in 'building back better' after a disaster is one thing, but much more important is to be proactive and invest now to build resilience against disasters yet to come – to learn to work much more effectively with the forces of nature than societies have done until now.

In this paper the authors argue the need to adopt integrated thinking and approaches into decision-making at the earliest possible moment. By doing this, there is an opportunity to consider the three pillars of sustainable development, and understand the options that exist and the trade-offs that may need to be made, for instance between economic efficiency and environmental sustainability, or between social equity and economic efficiency. The Global Water Partnership (GWP) works with partner organizations in many countries to develop such integrated approaches to water resources' management, to flood management and to drought management. In the run-up to the post-2015 negotiations, GWP has engaged stakeholders in 29 countries, through national-level processes. They have expressed clear policy messages for adopting integrated flood and drought management approaches through these recent consultations, details of which are included in this paper. As more and more people are affected by floods and droughts each year, and economic losses mount, there is much greater political urgency to prepare for these water-related disasters, given their increasing frequency and severity due to climate change (IPCC, 2007).

2. Floods and droughts as 'stress tests' for IWRM

The definition of IWRM is to develop and manage water, land and related resources in order to maximize the resultant economic and social welfare in an equitable manner without compromising the sustainability of vital ecosystems (Global Water Partnership (GWP), 2000). Clearly, too much or too little rain provides a stress test on the coordinated management of water and land resources. The process of IWRM is an adaptive approach that relies on coordinated decision-making across sectors and scales, involving many different stakeholders and institutions. Extreme deviations from normal conditions can be seen as a test of the robustness of the measures that societies are taking to manage water.

We look first at the integrated approach to flood management, which is firmly rooted in and takes its principles from IWRM, integrating land, water and risk management (see Figure 1). Integrated Flood Management (IFM) recognizes that floods can never be fully constrained, that they do not only pose a threat but also have beneficial impacts, and that these impacts affect many sectors of the economy. The concept shifts the focus from flood control to flood management and to flood risk reduction.

Any intervention to improve the use of the river basin assets (both in terms of water and land) must recognize that flood risk and its consequences (i.e. in human lives lost, and in economic loss and damage) cannot be totally controlled. The concept of IFM thus brings an innovation: although reducing loss of life should remain the top priority, the objective of flood loss reduction should be secondary to the overall goal of the optimal utilization of floodplains. It is only by taking the whole catchment area into consideration in the long term that the best use of scarce resources can be decided, and unintended consequences can be avoided. For instance, channelling flows for flood control upstream can result in increased flooding downstream; allowing development in the floodplain that seals off the soil surface and increases run-off also increases flood risks downstream. Straightforward benefit–cost ratio calculations for new infrastructure development, including flood control measures, need to give way to broader multi-criteria analysis, and both 'hard' interventions, such as dykes, and 'soft' interventions, such as early-warning systems, need to be in place. IFM aims to maximize the net benefits from floodplains, meanwhile minimizing loss of life. Such an integrated approach can be implemented only with the active involvement, which draws on the knowledge and skills, of all the affected stakeholders.

As the water resources within river basins are harnessed for social and economic development there is growing pressure placed on ecosystems, particularly aquatic and floodplain ecosystems. In many countries, this pressure is acknowledged with the emergence of environmental water reserves

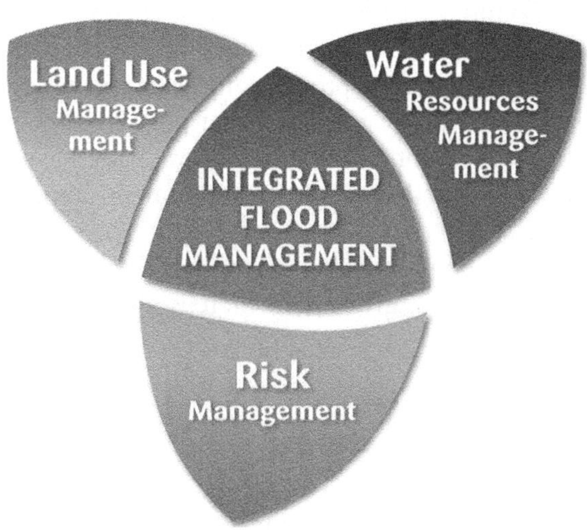

Fig. 1. IFM integrates land, water and risk management in river basins under the framework of Integrated Water Resources Management (World Meteorological Organization (WMO) & Global Water Partnership (GWP), 2013).

(Mexico and South Africa), water entitlements held for the environment (Australia), or environmental flows (Germany, Costa Rica). Infrastructure investments in many of these river basins have captured the surface run-off from small and medium flood events, and offered a degree of mitigation for large flood events. However, it is often these small and medium flood events that have ecological significance, especially in the riparian zone. Given this strong relationship between social, economic, and ecological dimensions of flood events, an approach based on IFM is best placed to assess the trade-offs that are being made within the overall river basin context.

The traditional approach to floods focused on reactive practices to reduce exposure to flooding and susceptibility to flood damage, mainly through structural measures separating the river from its floodplain. Over the course of time, these ad hoc interventions (often based on a narrow benefit–cost ratio analysis) proved to be only partially effective, shifting rather than mitigating flood risk. The acknowledgement of the necessity for a wider multi-disciplinary approach led to a paradigm shift from flood control to flood management and to flood risk reduction. This indicates a proactive mentality that recognizes that floods can never be fully constrained and that there will always be residual risks, but, at the same time, that floods also present positive aspects in terms of ecosystem benefits.

It is interesting to note that there is a growing awareness of the need for floodplain management from an integrated water and land-use perspective. In some rapidly developing countries this is leading to the identification of land-use zones that are designated to provide flood mitigation (as in China). In heavily developed floodplains such as in the Netherlands, a new approach is emerging that includes new infrastructure, breaching levee banks, recreating natural flood zones, and rethinking planning and development (Room for the River, 2007). Participatory processes, bringing to bear the views of a variety of stakeholders, are vital in order to enable these new approaches to succeed from a social perspective.

The defining feature of IFM, that of integration, is intended in both a horizontal and a vertical sense. While the first refers to a cross-sectoral decision-making process, the second entails a participatory and transparent approach at different levels of decision-making (national, regional, watershed scale, local). Together, they materialize in different forms: an appropriate mix of strategies, carefully selected points of interventions, and appropriate types of measures (structural or non-structural, short-term or long-term).

As a result, an IFM plan should be built on the following six pillars:

- manage the water cycle as a whole;
- integrate land and water management;
- assess and manage risk and uncertainty;
- adopt a best mix of strategies;
- ensure a participatory decision-making approach with a range of stakeholders;
- adopt integrated risk management approaches.

Integrated Drought Management – on the other side of climate extremes – also focuses on preparedness, prevention and mitigation. This takes place, however, in a very different context, since floods often appear rapidly, while droughts are slow-onset, creeping disasters that have a longer time frame for prediction and diagnosis. UN-Water (2013) notes that 'drought of different types (meteorological, hydrological, agriculture, socioeconomic) is a significant hazard and often results in catastrophic disasters. … drought impact is broad and data are scarce'. For many, a drought is a prolonged, abnormally dry period when there is not enough water for users' normal needs. However, as the Bureau of

Meteorology (BOM, 2011) notes that 'drought is not simply low rainfall ... Because people use water in so many different ways, there is no universal definition of drought.'

While the traditional approach to drought has often focused on an emergency, humanitarian response to hunger and loss of livelihoods, integrated drought management focuses on the whole cycle of disaster management, with a focus on the risk management aspect of the cycle as depicted in Figure 2. This longer-term approach enables integrated drought management to contribute to food security, poverty reduction and sustainable development efforts.

Traditionally, the response to drought has been reactive, managing the crisis rather than forecasting and taking preventive actions. This approach to drought management responds to the impacts of drought once they occur in an attempt to speed the recovery process. This crisis management approach has been noted to be costly, largely untimely, poorly coordinated, and often results in resources or assistance being misdirected. While the agricultural sector is often the hardest hit, drought can impact many sectors of the economy. One particularly crippling impact is on the generation of hydropower and other forms of energy, which then has a knock-on effect on social and economic activities across the spectrum – hospitals cannot function properly, traffic lights stop working, industries close down or work shorter hours.

Drought impacts illustrate the vulnerability of societies to drought. Assistance to those affected by drought, while essential in the short term, is treating the symptoms of vulnerability rather than its causes. In fact, many assistance programmes might result in increased vulnerability to future drought events by making individuals and societies more dependent on government programmes or assistance from donor organizations. As a consequence of an increased frequency of drought and societal vulnerability to extended periods of water shortage, the economic, social and environmental impacts of droughts have increased significantly worldwide. Because of their long-term socio-economic and environmental impacts, droughts are by far the most damaging of all natural disasters.

More recently, the focus in the immediate aftermath of a drought has shifted to building resilience, through planning and preparedness for the next drought. Local media coverage can be very useful in promoting the concept of drought-proofing, where the focus is on infrastructure for increased water

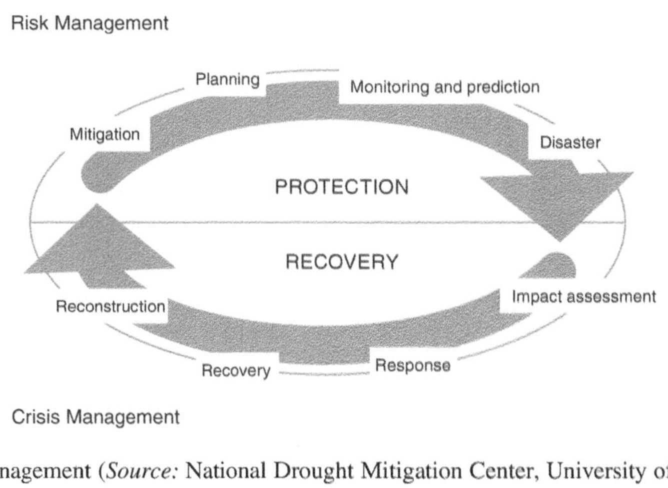

Fig. 2. Cycle of Disaster Management (*Source:* National Drought Mitigation Center, University of Nebraska-Lincoln in World Meteorological Organization (WMO) & Global Water Partnership (GWP), 2014).

storage capacity and the diversification of urban water supply options such as desalination, water reuse and recycling, and wastewater treatment. Energy supply also needs to be drought-proofed in various ways, while agriculture is often the hardest-hit sector of all.

A good example of an economic instrument that is starting to gain traction is water markets that can help water users and governments manage trade-offs between the different water uses as well as the environment, particularly in dry years (Grafton & Horne, 2014). In Australia the impacts of widespread drought in 2008–2009 saw the highest average prices for water (over A$400/megalitre (ML)), while 2010–2011, a relatively wet year, saw prices drop to A$20/ML. Such practices can be a market-based way of incorporating stakeholder perspectives – as these are then expressed through the water price. It must be noted, however, that a certain level of infrastructure and reticulation needs to be present to move water around in order for such economic instruments to be applicable.

Participatory approaches remain vital to finding solutions to drought challenges across various different economic sectors. For example, a central challenge that agricultural researchers are tackling is that of devising technologies that improve the drought resilience of agricultural production. Major food crops have been developed that are drought-tolerant or reach early maturity. This research relies on crops that have developed in harsh dry-land conditions by exploiting their drought-tolerance genes through plant breeding programmes. Scientists are either using molecular markers to better understand the genetic basis of drought tolerance and selecting for this trait or they are employing a practice called participatory plant breeding. The latter technique proactively involves farmers who observe productivity of the crops and select those plants that perform better under drought conditions. Despite some success stories, progress on building drought resilience in agriculture has been slow, and much remains to be done.[1]

An additional perspective is to look at the public-policy instruments that come into their own during a drought. Developing these with foresight requires sustained attention and considerable political will. Guidelines now exist to support countries in the development of a proactive drought policy and the various instruments to be applied (World Meteorological Organization (WMO) & Global Water Partnership (GWP) 2014). Experience in many countries shows that drought is a severe stress test for IWRM, as so many sectors are involved.

3. Establishment of APFM and work in countries since 2001

The Associated Programme on Flood Management (APFM) is a joint initiative of the World Meteorological Organization (WMO) and the Global Water Partnership (GWP) with the objective to promote the concept of IFM as a new approach to dealing and living with floods. To this aim, it facilitates dialogue and provides communities, civil-society organizations and government entities with guidance on flood management in order to apply an integrated approach to flood management. More specifically, the APFM's goals are to:

- promote the principles of IFM within the overall process of IWRM;
- identify gaps in present flood management practices, and stimulate partners to meet critical needs within their available human and financial resources;

[1] More information available at http://www.cgiar.org/www-archive/www.cgiar.org/pdf/drought_tolerant_crops_for_drylands.pdf.

- support IFM actions at all levels: national, regional, local and river basin-wide;
- provide a platform for a common strategic vision on IFM issues, and promote the implementation of effective policies and strategies worldwide;
- promote awareness about flood management issues, build political commitment and trigger action at all levels;
- provide advice and relevant information to institutions and decision-makers on flood management issues.

The APFM was founded in 2001 and has been implemented in three phases. The first phase (August 2001–July 2006) mainly focused on the definition of the basic principles of IFM (published in the IFM Concept Paper and the related Policy Series), the realization of regional pilot projects (in Africa in Kenya and Zambia; in Asia in Bangladesh, India and Nepal; and in Central and Eastern Europe in Bulgaria, Czech Republic, Lithuania, Poland, Romania, Slovak Republic and Slovenia), and the link to a global network of partners through GWP and WMO. The second phase (August 2006–March 2010) centred on the implementation of the concept of IFM through guidance, training and awareness-building at the local and regional level, and the establishment of the IFM HelpDesk. The third and current phase builds on previous work by strengthening the network of partners, providing long-term support and follow-up. At the same time, new activities are put in place, such as field demonstration projects, national and regional training courses, e-Learning and an enhanced communication strategy.

As shown in Figure 3, the HelpDesk provides, through the Get Help section, access to three different services:

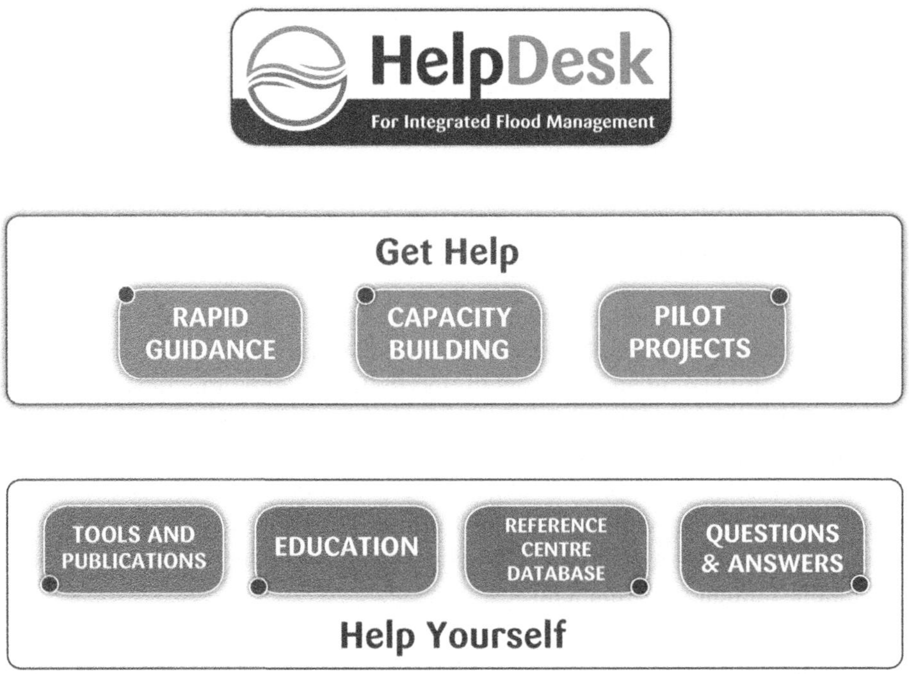

Fig. 3. The IFM HelpDesk (World Meteorological Organization (WMO) & Global Water Partnership (GWP), 2013).

1. Rapid Guidance, to get in touch with the APFM Technical Support Unit and obtain guidance on flood management policy, law and strategy;
2. Capacity Building, to request the organization of training courses tailored to the needs of the applicant;
3. Pilot Projects, to get assistance to implement IFM principles in the field, in cooperation with regional, national and local organizations.

Through the Help Yourself function users can access four sub-sections:

(1) Tools and Publications, from which it is possible to download the whole range of publications developed (the Policy Series, the Tool Series, Training Manuals, Case Studies);
(2) Education, a wide range of material to facilitate self-study targeting different kinds of readers (teachers and students, flood managers, policy-makers, trainers);
(3) Reference Centre Database, a set of interactive databases gathering details of institutions, policies and literature related to flood management;
(4) Questions & Answers, a collection of Frequently Asked Questions to redirect the user to the relevant section.

Although freely accessible online, the HelpDesk has been especially conceived as a demand-driven mechanism for use by government agencies in charge of flood management; by river basin organizations, bilateral and multilateral development agencies in the field of water resources and disaster management; by community-based organizations and non-governmental organizations (NGOs) engaged in flood preparedness; and by universities and other capacity-building institutions for the water and disaster management sectors.[2]

The HelpDesk is made possible by the active support of the Support Base Partners, a network of professional institutions and organizations contributing their expertise in IFM and technical backup in various areas of activity. These range from advice and advocacy in policy formulation, as well as technical issues, to facilitation of training courses on IFM, to development of tools and capacity-building materials.

4. Establishment of the integrated drought management programme and work in regions and countries since 2011

The approach to droughts has in the past frequently been captured by relief operations and a focus on emergency response, rather than an integrated approach that focuses on the aforementioned cycle of disaster management. The technology for drought early-warning systems is progressing very rapidly and is central to both approaches.

The US Agency for International Development (USAID) established the Famine Early Warning Systems Network (FEWS NET) to identify drought- and flood-related problems in the food supply system that potentially lead to famine or other food-insecure conditions in sub-Saharan Africa, Afghanistan, Central America, and Haiti. It combines remote sensing data and observations on the ground to provide early warning of critical situations to enable decision-makers to take action before an emergency arises.[3]

---

[2] The HelpDesk is fully operational and can be accessed online at http://www.apfm.info/?page_id=1253.

[3] More information available at http://www.fews.net/.

The WMO and the GWP established the Integrated Drought Management Programme (IDMP) in March 2013 at the High-level Meeting on National Drought Policy, to establish a long-term partnership approach and support countries in building resilience to droughts. The scope of the programme is to contribute to local, national and regional efforts for poverty alleviation in drought-affected regions of the world through an integrated approach to drought management cutting across sectoral, disciplinary, and institutional jurisdictions. It draws on the cycle of disaster management (Figure 2) – focusing on increasing coping capacity and resilience, rather than the traditional approach of crisis management.

Thus the IDMP aims to establish a culture of proactive drought management that focuses on preparedness and drought predictions, and on measures that mitigate the impact of droughts. The areas on which the IDMP is working are:

- better scientific understanding of, and inputs for, drought management;
- improved knowledge base, with better access to information and products;
- drought risk assessment, monitoring, prediction and early warning;
- policy and planning for drought preparedness and mitigation across sectors;
- drought risk reduction and response.

While the spatial scope is global, the results are expected to be policy-relevant and tailored to specific regional and national needs and requirements. The intended outcome of the IDMP is to support actors and partners in various sectors, disciplines and institutions to improve drought monitoring, prediction and preparedness on a global, regional, national and community scale, and to use this knowledge effectively in the development of short-term and long-term drought management plans and actions.

The GWP through its Regional Water Partnerships guides and assists in developing the regional linkages of the IDMP together with the WMOs and its links to HydroMet Services and its Regional Climate Centres. Four regional drought programmes have been established to date, building on the existing institutional capabilities and bringing in a partnering approach that engages multiple stakeholders. For example, the regional programme of the IDMP in Central and Eastern Europe (IDMP CEE) is led by GWP Central and Eastern Europe, collaborating closely with the Drought Management Centre for South Eastern Europe (DMCSEE).

The IDMP CEE develops guidelines for preparation of the drought management plans in ten countries, within river basin management plans according to the European Union Water Framework Directive. It also carries out national consultation dialogues to discuss preparation of drought management plans, compiles a compendium of good practices, establishes a drought information exchange platform, tests innovative solutions for improved drought resilience in six demonstration projects and organizes training and workshops.

The IDMP is gearing up to work in the Horn of Africa through its regional programme, IDMP HoA, and in West Africa through its regional programme, IDMP WAF, to add water management expertise to existing drought management efforts, and link with GWP's strong network of partner organizations on the ground in these regions. The programmes focus on identifying gaps through an integrated approach to drought management, bringing to bear the network of GWP through its country and regional water partnerships and WMO's expertise. Similarly, a Regional Drought Monitoring System is now being established across six South Asian countries.[4] All of these programmes are reliant on the development

[4] More information available at http://www.droughtmanagement.info/idmp-activities/south_asia/.

of ever-better and more-accurate climate services; hence, both the IDMP and the APFM form part of the Global Framework for Climate Services.

5. Flood and drought in the post-2015 development agenda

Flood and drought have been widely recognized as inhibitors of sustainable development, as they can wipe out development gains. For example, the Rio + 20 outcomes document, The Future We Want (UN, 2012), identifies '*the need to adopt measures to address floods, droughts and water scarcity*'. This perspective is also a central feature of the proposed SDGs being negotiated by the Open Working Group of the United Nations General Assembly.

As at 19 July 2014 the final proposals of the Open Working Group (UN, 2014) refer to disasters, including floods and droughts, under four separate focus areas (poverty, food security, cities, climate) as shown in Table 1. Unfortunately, while there was a specific target on water-related disasters under the global water goal (Goal 6) in earlier drafts, this was removed from the final proposals.

Between February and April 2014 the GWP brought together around 1,200 participants in 29 countries to give voice to stakeholders on proposals for a water goal and targets within the post-2015 development agenda. The consultations brought together voices from the environment, agriculture, planning, and infrastructure sectors, including political representatives, government officials, and delegates from the private

Table 1. Proposed Sustainable Development Goals and their associated targets where disasters are mentioned (as at July 2014).

Focus area	Proposed target
Goal 1: End poverty in all forms everywhere	Target 1.5: By 2030 build the resilience of the poor and those in vulnerable situations, and reduce their exposure and vulnerability to climate-related extreme events and other economic, social and environmental shocks and disasters
Goal 2: End hunger, achieve food security and improved nutrition, and promote sustainable agriculture	Target 2.4: By 2030 ensure sustainable food production systems and implement resilient agricultural practices that increase productivity and production, that help maintain ecosystems, that strengthen capacity for adaptation to climate change, extreme weather, drought, flooding and disasters, and progressively improve soil quality
Goal 11: Make cities and human settlements inclusive, safe, resilient and sustainable	Target 11.5: By 2030 significantly reduce the number of deaths and the number of affected people and decrease by [x] per cent the economic losses relative to GDP caused by disasters, including water-related disasters, with the focus on protecting the poor and people in vulnerable situations
	Target 11.b: By 2020, increase by [x] per cent the number of cities and human settlements adopting and implementing integrated policies and plans towards inclusion, resource efficiency, mitigation and adaptation to climate change, resilience to disasters, develop and implement in line with the forthcoming Hyogo Framework holistic disaster risk management at all levels
Goal 13: Take urgent action to combat climate change and its impacts	Target 13.1: Strengthen resilience and adaptive capacity to climate-related hazards and natural disasters in all countries

sector and civil society (see Box 1). These national consultations considered the proposals for a water goal and associated targets set out in the UN-Water Technical Advice paper (UN-Water, A Post-2015 Global Goal for Water: Synthesis and key findings and recommendations from UN-Water, 2014).

An overarching message that emerged from these national consultations (GWP, 2014) is the need for comprehensive and interrelated targets that further advance integrated approaches to water – 'Water is much more than a cross-cutting issue – unless the fundamental role of water and the water issues can be resolved, other important elements of the new development agenda will be unachievable.'

Climate change was also mentioned frequently throughout the consultations (e.g. Brazil, Bulgaria, and Tajikistan) with regard to a water goal. Participants highlighted that the impacts of climate change are already visible worldwide with the growing intensity and unpredictability of floods and droughts. Looking to the future, the participants argue that there are clearly strong links between adaptation to climate change and addressing the requirement of a 'secure' and 'sustainable' source of water for all. Therefore, a water goal with a target that focuses on water-related disasters is considered crucial.

At national level there are also calls for a water-related disasters target to give more emphasis to drought, as experience suggests that there is more attention given to floods. The consultations also offer some insight into the gaps that are perceived to exist in our ability to address water-related disasters. For example, there are calls for greater knowledge about water-related disasters; implementation of early-warning systems, resilience and drought-proofing; greater community participation; greater focus on planning and preparedness rather than on emergency response; and improvements in underlying data and monitoring systems.

At a global level, through the SDG process, there is a very active debate on the whole question of how countries might approach the integrated management of drought and flood, within the broader context of sustainable development. In doing so, there is not really any concern about the language or public perceptions – this approach is seen as a self-evident public good. However, in situations where there are highly developed land and water rights, strongly connected to economic production, a different debate emerges.

In Australia, for example, the public discourse has moved on from DRR related to droughts and floods, to a new concern for the environment (see Box 2). From a technical perspective, it is widely accepted that regulation of river flows in the Murray–Darling Basin has modified natural flow regimes (MDBA, 2011). The consequence of this, combined with high levels of extraction, has been a steady decline in the environmental health of the basin. With recent reforms there is now significant effort to restore the environmental health of rivers, wetlands, and floodplains – known as environmental watering.

However, it is interesting to consider the public reaction to this and the emergence of a new language for environmental watering. Hydrologically, environmental watering is the manipulation of river flows to raise river level to create defined flood events at specific locations and times. Typically, these events would mimic the small- and medium-sized flood events that would have occurred prior to river regulation and extensive storage of surface run-off. In the Australian example a clear distinction is being drawn between 'overbank flows', 'natural flow regimes', and the term 'flooding', which invariably creates a negative public reaction. There is always a residual risk that needs to be carefully managed, in terms of expectations and reactions. Issues of compensation and blame arise especially where river managers must manipulate river operations to generate the overbank flows – in other words, the flood is being created; therefore, the flooding does not fall into the general category of 'acts of God' or 'force majeure'. Under these circumstances flood easements or other mechanisms may be necessary to allow the delivery of 'overbank flows'.

Box 1. Perspectives on a water goal and its implementation, expressed during national stakeholder consultations held in 2014 (GWP, 2014)

Argentina: '*It is suggested that in the basis of the target more emphasis is made on droughts, since our experience shows that they are not given the same attention as floods*' ... '*Since Argentina is also very affected by climate change ... both with respect to floods and droughts, the inclusion of this target is positively valued*' ... '*Mortality reduction and economic loss strongly depend on the progress that can be made in the other sustainable development goals*'.

Bangladesh: '*About 230 natural disasters occurred in the country between 1980 and 2010, causing death to 190,000 people, livelihood loss of US$320 million and economic loss of US$550 million. In terms of death, the cyclonic storm is the most devastating disaster that caused 87% of total deaths. And in terms of livelihood and economy, flood is the worst disaster in terms of the number of affected people*'.

Bulgaria: '*Nowadays almost no flood protection investments are made, only emergency repairs and breaks are funded to protect endangered settlements and valuable agricultural land*'.

Colombia: '*Colombia has advanced towards a risk management scheme that has improved the understanding of disasters, the credibility of public entities with communities, and therefore mitigates, to a certain point, the impact of disasters. However there is still a long way to go*' ... '*Reduction of disaster impact and the usefulness of early warning systems require big doses of community participation and capacity building, not only socialisation*'.

Ghana: '*On management of floods, the meeting proposed the development and implementation of drainage master plans for all communities at risk, backed by implementable disaster management plans*'.

Guatemala: '*The vulnerability of the country to extreme events had generated the need to develop a national strategy of risk management, which goes beyond the theme of water*' ... '*Investment is required in development of land management municipal plans, with the risk management component*'.

Indonesia: '*Ecosystems can provide services for drinking water supply, water for food production, wastewater treatment, and disaster risk reduction*'.

Nicaragua: '*It is necessary to build the capacity of government institutions in order to establish mechanisms to evaluate Early Warning Systems (EWS) at the national level and identify communities with high levels of risk*' ... '*it is also necessary to provide training spaces to these communities on early warning systems created from locally available/low cost materials, disaster risk reduction and community-level contingency planning*'.

Pakistan: '*droughts show repeat visits with both prolonged and seasonal occurrences*' ... '*disaster risk reduction activities need to be sustained during peace time and stock building made a regular feature to deal with upcoming challenges of drought*'.

Peru: '*In Peru, the water-related disasters have been increasing in the last 30 years*' ... '*social networking of stakeholders committed to disaster risk management, including young people, women, authorities, and others will help in the generation of resilience capabilities*'.

Poland: '*Poland is a country suffering from droughts in the vegetation growing season, which has a negative impact on water management, especially in agriculture*'.

Tanzania: '*It is important to note that disaster risk management is a multi-sector and multi-stakeholder issue which engages many relevant ministries and institutions*'.

Zambia: '*On a yearly basis about 19 districts experience floods while 13 districts usually experience droughts*'.

Box 2. Droughts and floods in Australia

The 'Millennium Drought' (2001–2009) can be described as the worst drought on record for south-east Australia (van Dijk *et al.*, 2013), with consequences for ecosystems, the economy, and society. Particularly severely hit by the Millennium Drought were river ecosystems and irrigated and dry-land agriculture in Victoria and the Murray–Darling Basin (MDB), Australia's largest river system. During this period river regulation provided a highly effective means to manipulate and distribute the impacts of drought on riverine ecosystems and communities, mitigating the impacts for some by exacerbating them for others. Water-sharing rules and trading favoured users reliant on continued supply, including users with needs for irrigation of perennial high-value crops, and users such as the water utilities, while lower-security water users and riverine ecosystems bore the consequences.

In Australia drought policy can be traced back to the early 1900s. Over the last century the focus of support has changed over time, from drought-proofing, through irrigation, to direct financial assistance. With the adoption of the 1992 National Drought Policy, the emphasis shifted again from drought being classified as a natural disaster towards that of its being a normal component of the operating environment – a predictable risk that should be carried by those directly impacted (ABARES, 2011).

However, while drought has been a significant driver of water policy in Australia over the last century, the issue of floods has also been significant. As a land of 'drought and flooding rains', Australia has been at the forefront of infrastructure development to capture surface water run-off where practicable to provide inter-annual storage. This, combined with infrastructure to regulate flow in rivers, underpinned the expansion of irrigation during the latter decades of the twentieth century. Given the history of economic development based on harnessing water resources, the complex nature of droughts and floods, and the consequential opportunity for cumulative impacts, interactions and feedbacks, there has also been a gradual change and broadening in public discourse in Australia to emphasise 'natural' environmental processes.

The Australian example provides two salutary messages for river basins in other continents that are currently subject to rapid development. Firstly, it becomes very difficult and expensive to rein back water over-allocations where the institutional mechanisms are not sufficiently flexible. Secondly, an integrated perspective of water resources' management from the earliest stages represents the best mechanism to balance the social, economic, and environmental interests in water resources, and thereby avoid the future costs in redressing the balance.

6. Conclusions

From a long-term perspective on droughts and floods, there are clear differences in priorities between highly developed and regulated river basins and more-natural and undeveloped river systems. Early years of investment are heavily dominated by the urgency to regulate rivers for drought-proofing as well as to mitigate flood events, and to harness the resource base for social and economic benefit. It is clear that the parameters of this debate (floods are bad) live long in the public psyche and through the cumulative impacts of ever-more-sophisticated policy and operations. The net result is one where society loses sight of the inherent ecosystem values that floods have, and the fact that these need to be integrated back into river basin development at the earliest possible opportunity. Where these ecosystem values are not integrated from the outset, the long-term cost means high levels of public investment to restore river flows for ecological benefit and for the resultant socio-economic benefit.

With climate change as an ever-more-present reality, whose impacts are most often experienced through water-related disasters such as floods and droughts, there is an urgent need to build resilient societies through more integrated policies and practices, including stakeholders' perspectives and a partnership approach.

- Strengthening water security requires a proactive and integrated approach to water-related disaster management, recognizing that the changing socio-economic conditions and levels of development bring about new expectations of what constitutes water security.
- As river basin management develops in terms of infrastructure, operations and policy, social perspectives on drought and flood change over time.
- Starting with an integrating approach in mind is the best way to avoid long-term costs and anthropogenic damage to the natural functioning of river basins.
- An integrated approach to flood management involves looking also at the benefits of floods.
- Integrated drought and flood management needs to consider the whole cycle of disaster management as shown in Figure 2, aiming to shift the focus away from short-term responses towards longer-term preparedness, prediction and prevention, with practical measures that reduce the impact of these extreme events.
- Evidence from many countries shows that there is a popular groundswell of support for better water security, and an increased awareness of water-related risks. As expressed by the stakeholders quoted in this paper, a target on water-related disasters within the SDGs is crucial, to ensure that droughts and floods attract the necessary political will, focus and investment.

References

ABARES (2011). *Drought in Australia: Context, Policy and Management.* Australian Bureau of Agricultural and Resource Economics and Sciences, Canberra.

BOM (2011). *Living with Drought.* Bureau of Meteorology, Canberra.

Global Water Partnership (GWP) (2000). Integrated Water Resources Management, Global Water Partnership Technical Advisory Committee (TAC) Background Paper No. 4, available at: http://www.gwp.org/Global/GWP-CACENA_Files/en/pdf/tec04.pdf.

Global Water Partnership (GWP) (2014). *National Stakeholder Perspectives on a Water Goal and Its Implementation.* Global Water Partnership, Stockholm.

Grafton, R. Q. & Horne, J. (2014). Water Markets in the Murray–Darling Basin. In: *Global Water: Issues and Insights.* Grafton, R. Q., Wyrwoll, P., White, C. & Allendes, D. (eds). ANU Press, Canberra, pp. 37–48.

IPCC (2007). Fourth Assessment Report (AR4): Climate Change, available at: http://www.ipcc.ch/publications_and_data/publications_ipcc_fourth_assessment_report_synthesis_report.htm.

MDBA (2011). *The Proposed 'Environmentally Sustainable Level of Take' for Surface Water of the Murray–Darling Basin.* Murray–Darling Basin Authority, Canberra.

Room for the River (2007). Approved Decision Room for the River, available at: http://www.ruimtevoorderivier.nl/english/room-for-the-river-programme.

UN (2002). Plan of Implementation of the World Summit on Sustainable Development, available at: http://www.un.org/esa/sustdev/documents/WSSD_POI_PD/English/WSSD_PlanImpl.pdf.

UN (2012). *The Future we Want*. United Nations, New York.

UN (2014). Report of the Open Working Group of the General Assembly on Sustainable Development Goals (Document A/68/970).

UNCCD (2007). The 10-year strategic plan and framework to enhance the implementation of the Convention, available at: http://www.unccd.int/Lists/SiteDocumentLibrary/10YearStrategy/Decision%203COP8%20adoption%20of%20The%20Strategy.pdf.

UN ISDR (2007). Hyogo Framework for Action 2005–2015: Building the Resilience of Nations and Communities to Disasters Extract from the final report of the World Conference on Disaster Reduction (A/CONF.206/6), available at: http://www.unisdr.org/files/1037_hyogoframeworkforactionenglish.pdf.

UN-Water (2013). UN-Water Country Briefs: Project Description of the method used, available at: http://www.unwater.org/fileadmin/user_upload/unwater_newdocs.

UN-Water (2014). A Post-2015 Global Goal for Water: Synthesis and key findings and recommendations from UN-Water, available at: http://www.unwater.org/fileadmin/user_upload/unwater_newdocs-Water_paper_on_a_Post-2015_Global_ Goal_for_Water.pdf.

van Dijk, A. I. J. M., Beck, H. E., Crosbie, R. S., de Jeu, R. A. M., Liu, Y. Y., Podger, G. M., Timbal, B. & Viney, N. R. (2013). The Millennium Drought in southeast Australia (2001–2009): Natural and human causes and implications for water resources, ecosystems, economy, and society. *Water Resources Research* 49, 1040–1057.

World Meteorological Organization (WMO) (2009). WCC-3 High Level Declaration, available at: http://www.wmo.int/wcc3/declaration_en.php.

World Meteorological Organization (WMO) & Global Water Partnership (GWP) (2013). Associated Programme on Flood Management brochure, available at: http://www.apfm.info/about/APFM_HelpDesk_Flyer_2013.pdf.

World Meteorological Organization (WMO) & Global Water Partnership (GWP) (2014). *National Drought Management Policy Guidelines: A Template for Action* (D. A. Wilhite). Integrated Drought Management Programme (IDMP) Tools and Guidelines Series 1. WMO, Geneva, Switzerland and GWP, Stockholm, Sweden.

Integrated actions in the management of critical hydrologic events

Bruna Craveiro De Sá E Mendonça* and Mateus Monteiro De Abreu

Agência Nacional de Águas (ANA), Setor policial (SPO), Área 5, Quadra 3, Bloco M. 70610-200, Brasília, Brazil
**Corresponding author. E-mail: bruna.mendonca@ana.gov.br*

Abstract

Owing to the increase in urban populations, there has also been an increased incidence of critical events, especially those related to floods and landslides. Since the implementation of the National Civil Defense System in Brazil, a focus on disaster management is no longer the answer when assisting those affected; instead, the answer has become managing disaster risk, in addition to the response, including prevention and minimizing the effects of the critical event. This paper highlights the actions for modernization and expansion of the hydrological monitoring network, in particular the alert network, the completion of an atlas of vulnerability to floods and the deployment of Situation Rooms in the states, with a view to strengthening the role of these bodies in disaster risk management, and enabling greater responsiveness to these events and the use of the information gathered as a water resources management tool, allowing for better use of public resources.

Keywords: Critical hydrologic events; Flood vulnerability; Monitoring; Risk management; Situation Rooms

1. Introduction

Brazil's National Water Agency (Agência Nacional de Águas; ANA), created by Law No. 9984 on 17 July 2000, is an independent national regulatory body, with administrative and financial autonomy, under the Ministry of Environment. ANA is part of the National Water Resources Management System and is responsible for implementing the National Water Resources Policy which has, among its objectives, the prevention of and protection against critical hydrological events (droughts and floods) that occur naturally or as a result of inappropriate use of natural resources.

ANA has the following among its duties:

- Planning and promoting actions to prevent or minimize the effects of droughts and floods, within the National Water Resources Management System, in conjunction with the National Civil Defense and Protection System, to support the states and municipalities. These actions, when they involve the

doi: 10.2166/wp.2015.010

application of preventive rationing, may only be promoted observing criteria defined by federal decree, after consulting the respective river basin committees, if any.

- Setting and monitoring reservoir operation rules to be observed by public and private actors, in order to ensure multi-purpose water uses, according to the river basin master plans. The definition of the operation rules for the reservoirs of hydropower plants is to be carried out in conjunction with the Operador Nacional do Sistema Elétrico (ONS; the Electric System National Operator).
- Promoting the coordination of activities under the national hydrometeorological network in cooperation with agencies and with public or private bodies that comprise or are users of it.

ANA started to operate its Situation Room in November 2009. Through this facility, the agency monitors the hydrometeorological conditions of priority river basins and of the main reservoirs of the country, in order to support, in particular, decision-making to minimize the effects of droughts and floods. To do so, data from monitoring rainfall levels and river flows, the operation of the main reservoirs, weather forecasts and climate are used, as well as hydrological models and records of occurrences of emergency, or states of emergency, in the municipalities.

From the flood events which occurred in June 2010 in the states of Alagoas and Pernambuco, in the Mundaú, Paraíba, Una and Sirinhaém Capibaribe river basins, which resulted in the loss of human lives and properties as well as dozens of displaced and thousands of unsheltered families, the Agency decided to support the states in structuring their own situation rooms.

Similarly to ANA's, the State Situation Rooms carry out monitoring, but on a different spatial scale of analysis. The situation rooms act as a management center and subsidize the water resources state manager in critical situation decision making, identifying the possible occurrences of critical events (droughts and floods) by monitoring the hydrological conditions of the major river systems of the state and the meteorological data. Thus, they allow the adoption of preventive measures to mitigate the effects of droughts and floods.

To provide the State Situation Rooms with information, especially about floods, it was necessary to know the areas subject to floods and to verify the impact level and frequency of occurrence of those events.

The Flood Vulnerability Atlas was designed as a diagnostic tool, which resulted in the obtaining of information on the frequency of occurrence of floods and the impact levels associated with them. The intersection of these two pieces of information generated the classification of river stretches according to their vulnerability to flooding, on a scale of 1 to 1 million, in all Brazilian states, which allowed the definition of the warning hydrological network in the states.

2. The role of the National Water Agency

2.1. ANA Situation Room

Owing to its geological, geographical and climatological characteristics, floods, droughts, and landslides are the most common natural disasters in Brazil, and are strongly related to the occurrence of weather phenomena, in particular so-called 'extreme events'.

Floods and droughts have increasingly drawn the attention of society, since they cause significant economic and social impacts. The year of 2009, in particular, was marked by the occurrence of such events resulting in significant damages.

That same year, the ANA Situation Room was inaugurated with the basic function of monitoring the hydrological trends across the country. This task is accomplished by analyzing the evolution of rainfall levels and flow rates of rivers and reservoirs, and the weather and climate, as well as performing mathematical simulations that help in predicting extreme events (droughts and floods) across the country, in accordance with the assignments given to the Agency by Law No. 9.984/2000.

Operated jointly by the Superintendency for the Management of the Hydrometeorological Network and the Superintendency of Multiple Uses and Critical Events, the Situation Room performs the collection and validation of data and its analysis, aimed at producing reliable and timely information for decision-making by the ANA Board of Directors.

2.2. Process of coordination with federal agencies

Actions aimed at prevention of critical hydrological events carried out by ANA are part of a set of actions undertaken at the federal level, in the field of risk management and disaster response. In recent years, a growing concern regarding risk identification and prevention of natural disasters has been observed in Brazil, replacing the traditional treatment given to the subject, which focused predominantly on disaster response.

In this sense, institutions dedicated to gathering and networking relevant information to cope with extreme events were created, notably the National Center for Monitoring and Early Warning of Natural Disasters (CEMADEN) and the National Center for Risk and Disaster Management (CENAD). CEMADEN collects and produces information and systems for monitoring and alerts concerning the occurrence of natural disasters in susceptible areas throughout Brazil, while CENAD aims to manage strategic actions and disaster response, as shown in Figure 1. Within this structure, CEMADEN sends alerts of possible occurrences of disasters in mapped risk areas to CENAD. CENAD, in turn, transmits alerts to states, municipalities and other federal agencies and supports the disaster response actions.

In August 2012, the National Risk Management and Disaster Response Plan (Brazil Ministry of Planning, 2012) was launched with the goals of protecting lives, ensuring safety and minimizing damage resulting from disasters, and preserving the environment. The Plan articulates actions of different institutions, divided into four thematic axes – prevention, mapping, monitoring and alert, and disaster response.

2.2.1. Prevention axis. Prevention includes the infrastructure built as a result of the growth acceleration program (PAC) aimed at reducing the risk of natural disasters. Among the actions in this axis are slope stabilization work, urban drainage and flood control, and the construction of withdrawal, delivery and storage systems for drinking water in semi-arid regions to face the effects of drought.

2.2.2. Mapping axis. This axis provides the mapping of areas at high risk of slippage, landslides and floods in 821 priority municipalities. In these municipalities, intervention plans (which identify the vulnerability of houses and infrastructure within the risk areas and propose solutions to the problems encountered) will be prepared, in addition to supporting the preparation of geotechnical maps of urban fitness, to assist the municipalities in spatial planning. The development of the Flood Vulnerability Atlas is included in the 'hydrological risk' component.

2.2.3. Monitoring and alert axis. The actions under this axis aim to strength the monitoring and alert system, especially through the expansion of the observation network and the structuring of CEMADEN and CENAD. It also covers the implementation of the State Situation Rooms for hydrologic monitoring.

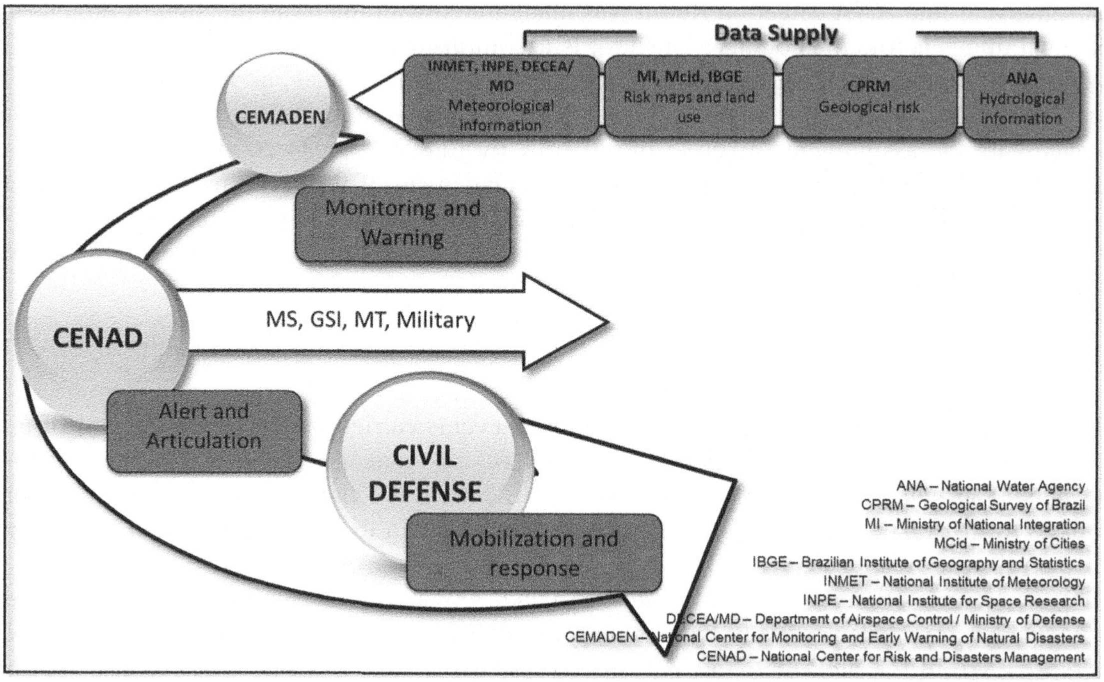

Fig. 1. Cycle of risk management and response to natural disasters.

2.2.4. Disaster response axis. This axis involves a set of actions which aim to increase the responsiveness of disasters occurrences, such as the creation of the National Emergency Force and the mobilization of National Security Forces to support the states and municipalities when major disasters occur, aiming to accelerate the implementation of recovery and relief actions.

The main role of ANA in this system, in accordance with Law No. 12,608, of 10 April 2012, is to continuously produce and forward to CEMADEN and CENAD reliable hydrological information with appropriate frequency and sufficiently in advance to allow adequate decision-making in a timely manner. In the case of the occurrence of critical flood events, a task force of geologists and hydrologists is temporarily mobilized in order to follow the event more closely. In parallel, ANA has elaborated its Flood Vulnerability Atlas, designed as a diagnostic tool of occurrence and impacts of gradual flooding in major rivers of the Brazilian basins. This project consisted of the identification of reaches of rivers where gradual flooding or plain flood occurs, the evaluation of the vulnerability of the affected regions and the definition of critical areas. Flood vulnerability maps have been made for each of the Brazilian states.

2.3. Process of coordination with the states

Following the flood events which occurred in June 2010 in the states of Alagoas and Pernambuco which resulted in the loss of human lives and material goods, as well as the displacement and a lack of shelter for tens of thousands of families, the agency decided to support the States in structuring their own Situation Rooms, following the existing ANA model and integrated with it.

These rooms are centers for managing critical situations related to water, with the aim of identifying possible occurrences of critical events and thus allowing the adoption of preventive and mitigatory measures that minimize the effects of droughts and floods. In addition to the state water resources management agency, State Situation Rooms typically rely on the presence of the local technical meteorological institute and the State Civil Defense. The working scale and the local knowledge gathered there allow the detection of and attention to regional events, unlike in ANA's Situation Room, which operates nationwide and on a macro scale.

Through technical cooperation agreements between ANA and the states, signed to set up the State Situation Rooms, the agency provides the necessary infrastructure for the rooms and data collection platforms (DCPs), to compose a monitoring and alert network in the major watersheds affected by floods, identified in the Flood Vulnerability Map. It also provides training and necessary software for their operation. The states, in turn, are committed to providing the physical space and furniture to deploy the rooms, as well as specific technical staff to perform the office and field activities necessary for their proper functioning.

Each technical cooperation agreement requires an annual basis work plan which provides the implementation of activities that contribute to the process of deployment and operation of the rooms and their integration with the Situation Rooms of ANA and other federal, state and municipal bodies. So far, ANA has helped establish Situation Rooms in all of the country's state capitals and in the Federal District.

In 2012, the program designed to support the implementation of Situation Rooms joined the monitoring and alert axis of the National Risk Management and Disaster Response Plan, which had its implementation monitored by the Casa Civil (Office of the Chief of Staff of the Presidency). Moreover, CEMADEN will receive reports from the Situation Rooms and send notices to CENAD, following the example of interaction achieved between CENAD and the Pernambuco Situation Room.

The Flood Vulnerability Atlas also had the participation of states in its preparation, which assisted in the identification of vulnerable river stretches and in the estimation of the frequency and impact of gradual floods. From this information, the vulnerability of river stretches and critical basins within a state were defined. This study contributed to an indication of the need and location for additional telemetric hydrometeorological stations, which are monitored and maintained by State Situation Room staff.

In the long term, ANA will support states in developing flood hazard maps, flood risk maps, alert levels in rivers and reservoirs and the impact of dam breaks. Furthermore, ANA also supports the development or improvement of hydrological forecasting systems.

2.3.1. Spatial distribution of critical events related to water. First, it is important to note that the phenomena of drought and flood are distinguished in several ways: while floods affect cities located on the margin of rivers, hydrological droughts affect broader regions where there is insufficient water to meet demand. Moreover, floods generally are processed much more quickly than droughts, which are usually recorded after long periods of a negative precipitation anomaly. On the other hand, floods are usually associated with high rainfall sufficiently able to raise the level of a river beyond the limit supported by its channel, natural or artificial, which shows a close relationship between a weather event and urban occupation and the occurrence of a flooding event.

To synthesize the distribution throughout the Brazilian territory of critical hydrological events of floods and droughts monitored in ANA's Situation Room, the next items considered were the main results presented in the Brazilian Atlas of Natural Disaster (Brazil Ministry of National Integration & Federal University of Santa Catarina, 2012) published by the Ministry of National Integration, and shown in the maps and charts which follow.

2.3.1.1. Floods

The term 'flood' can be understood as the overflow of water from the normal channel of rivers, seas, lakes and reservoirs, or water accumulation by poor drainage in areas not normally submerged. Depending on their magnitude, floods are classified as exceptional, of high magnitude, normal or regular, or of small magnitude.

Gradual flooding includes floods where the gradual rise of the water level of a river, above its natural channel, occurs. The prediction of the occurrence of such events can be made with the use of ANA's river flows monitoring network. Thus, the monitoring developed in the Situation Room is more focused on the prediction of gradual flooding. To assist in understanding how such events are distributed over the Brazilian territory, they are presented below in Figures 2 and 3.

An analysis of Figure 3 shows that the southeast region has the highest number of recorded cases of gradual flooding in the country, followed by the northeast and south regions. Considering monthly distribution, there is a highlight in the peak occurrences in the southeast in January. The high number of occurrences in the north and northeast in April also stand out.

2.3.1.2. Flood Vulnerability Atlas

This Atlas is focused on gradual flooding or plain flooding events, which have the slow ascent and descent of river levels as their main characteristic. These events are seasonal almost all over the country, with the beginning of floods varying depending on the rainy season (Figure 4). In the basins of the

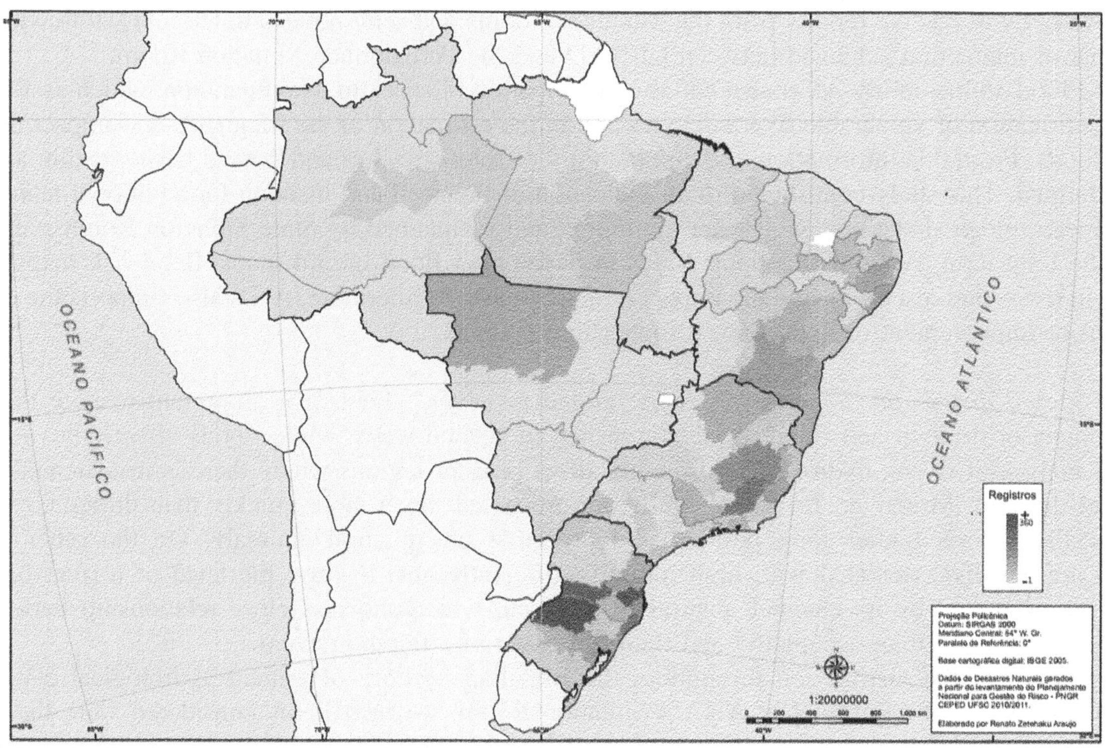

Fig. 2. Natural disasters caused by gradual flood in Brazil from 1991 to 2010 (Brazil Ministry of National Integration & Federal University of Santa Catarina, 2012).

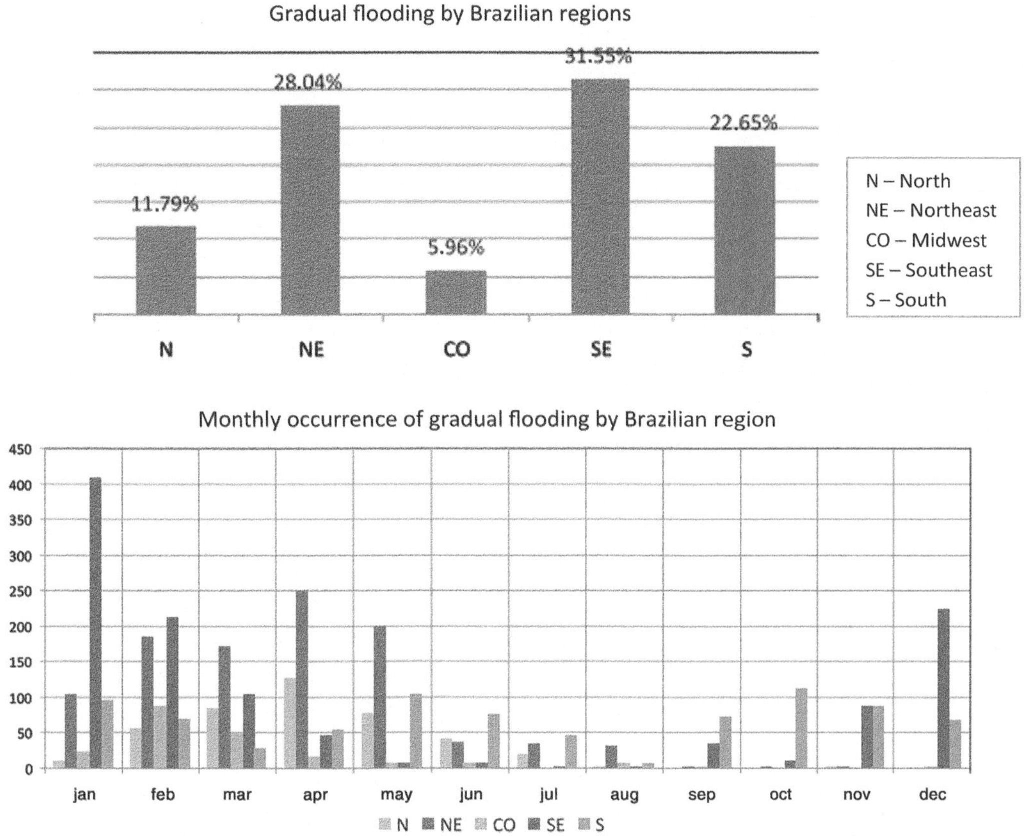

Fig. 3. Distribution of disaster caused by gradual floods in Brazil by region for the period from 1991 to 2010 (top) and the monthly occurrence of gradual flooding by region (bottom) (Brazil Ministry of National Integration & Federal University of Santa Catarina, 2012).

southeast, northeast and south regions, the flooding cycle is well defined, and the events generally occur at intervals of days or weeks and can repeat successively throughout the rainy season. In the north and midwest, large basins such as the Amazon and the Pantanal, have long cycles of floods, which last for months. They are also disasters with high economic loss in the affected areas, although the number of deaths is lower than in other water-related phenomena, such as torrents during heavy rains.

The flooding information was gathered in each state through meetings held during 2011 and 2012, in which the most relevant actors in tackling the issue were put together. All meetings were conducted by the state water resources management agency, with the participation of ANA. The involvement of the State Civil Defense is also highlighted. In each state there was a different composition of the participants, considering their involvement in the theme, region and the organizational structure of the State Government. The Geological Survey of Brazil (CPRM), the Hydroelectric Company of San Francisco (CHESF) and the National Department of Works Against Droughts (DNOCS)[1] also attended the meetings.

[1] CHESF and DNOCS attended the meetings held in their area of activity (basically, the Brazilian northeast).

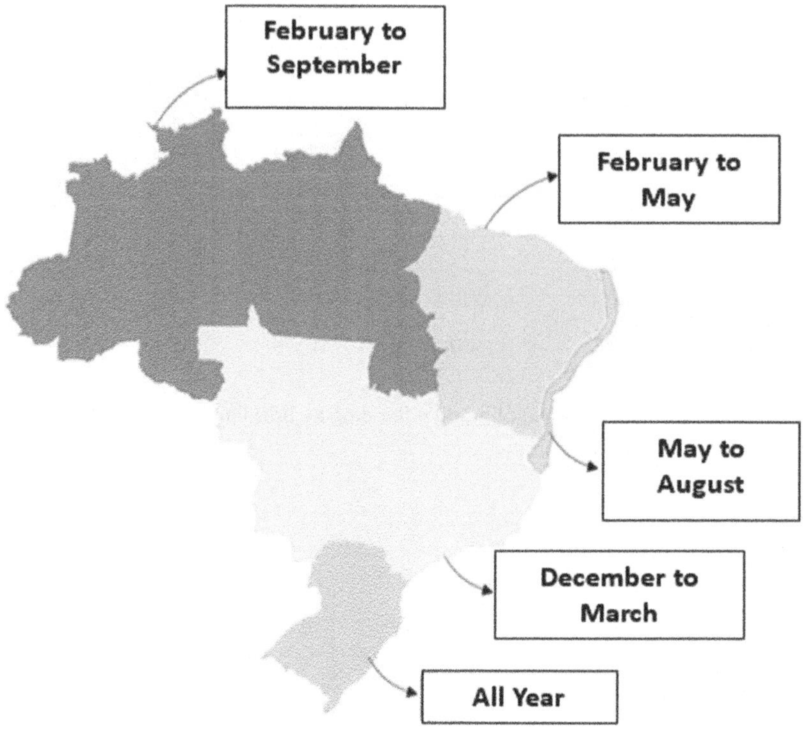

Fig. 4. Critical periods for flood monitoring in the Brazilian regions.

To obtain the effective participation of states in preparing the maps, the first activities were the presentation of the objectives and methodology of the Atlas, through meetings in the states, which primarily involved the water resource management bodies and civil defense organizations. On these occasions, ANA presented preliminary maps from which each state could meet its set of information and records of flood events. Some states chose to hold several meetings in addition to the events organized in conjunction with ANA, as well as workshops to review the maps, which extended the deadline initially set.

The project was developed following the steps listed below, which have their own methodology and products:

(a) flooding stretches;
(b) vulnerability of flooding stretches;
(c) critical areas.

2.3.1.2.1. Flooding stretches. The purpose of this step was to identify the geographical distribution of the occurrences of floods in Brazil's major rivers, their frequency, magnitude and associated impacts, i.e. to gain a picture of events occurring in a given length of time.

The flooding stretches were identified by visits to the water resource management agencies and the states and the Federal District Civil Defense, in which the following issues were evaluated:

- where floods occur in the state;

- how often the river stretch flooded;
- what is the impact caused by flood in each stretch.

At this stage, a map of flooding stretches was consolidated for major rivers on a scale of 1 to 1 million, gathering the information collected in the states. The municipality boundaries and the municipality headquarters were also represented on the map in order to serve as a physical basis for signaling the flooding stretches. When working at the national level, the issue of scale becomes a relevant factor for the representation of information. The waterbase used today by ANA for application throughout Brazil is the so called 'Hintegrada', which is comprised of the graphical representation of the systematic mapping rivers at the scale of 1 to 1 million, as required by the Brazilian Institute of Geography and Statistics (IBGE), using the Otto Pfafstetter coding methodology. From ANA's basic hydrological data, a common period of daily rainfall data was determined for selected rainfall stations, to perform frequency analysis in order to identify the most recurrent extreme daily values, from which the occurrence of floods was estimated.

Even as a by-product, a map based on 'Otto' sub-basins was generated, which is useful for many other purposes at ANA, such as studies of grants of water, hydrologic modeling, identification of drainage areas of hydrological stations and alert systems.

2.3.1.2.2. Vulnerability of flooding stretches. The purpose of this step was to estimate the flooding stretches vulnerability in relation to the frequency of floods and their impacts. The vulnerability results from a qualitative assessment of the data gathered in the previous step, undertaken by classifying the frequency of occurrence and the level of impacts associated with flooding.

The frequency of floods is classified in the following ranges:

- High: floods occur within less than 5-year intervals;
- Average: floods occur between 5- and 10-year intervals;
- Low: floods only occur within intervals exceeding 10 years.

The level of impact was rated according to the criteria below (Figure 5):

Finally, the vulnerability was evaluated by performing the following combination of frequency and impact: high, when the impact is high for any frequency or when the impact is medium and the

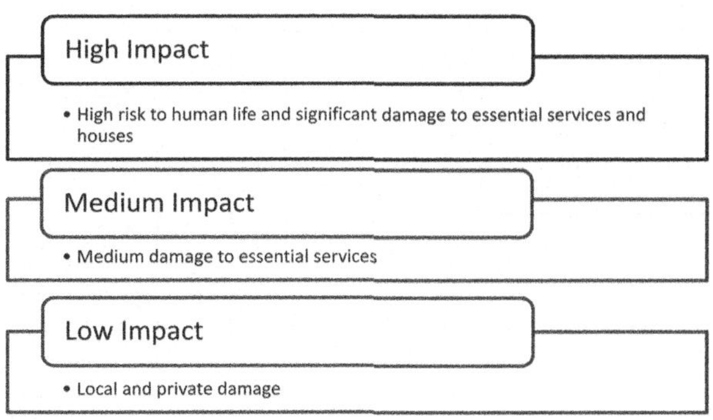

Fig. 5. Classes of impacts.

frequency is high; low when the impact is low and the frequency is medium or low; medium in all other cases, as shown in Figure 6.

The Annual Flood Occurrence Map serves to validate the results obtained in this step. Records of the occurrence of flooding and consequent damage to the population were determined by the National Secretariat of Civil Defense (SEDEC/MI) and IBGE, which were compiled and stored in a georeferenced database for application in the 1:1000000 scale.

The most significant disasters and areas of greatest frequency were identified and ranked by municipality, following the form of organization by the Civil Defense and IBGE. The National Secretariat of Civil Defense provides, in addition to recording when and where the event occurs, data on damage to people and property, by consulting reports submitted by the municipalities. IBGE has recently published the results of the National Survey on Sanitation (2008), in which the incidence of flooding in cities over the past 5 years was investigated. As a result it was found that 48.7% had no flood problems, 27.4% had floods or overflow problems and 23.9% had flooding or overflow issues due to bottlenecks. Also it was discovered that 60.7% of the municipalities had urban occupation in areas that are naturally subjected to flooding.

After the water management bodies and the states and Federal District Civil Defenses reviewed the results, the maps were drawn from the IBGE Continuous Cartographic Base of Brazil (BCIM) to the Millionth Scale. The hydrography follows the Otto Pfafstetter encoding to identify each river stretch, seeking their integration within the National Water Resources Information System (SNIRH). All information feeds a geographic information system, from which the maps are developed for each theme. Figure 7 summarizes the steps taken to build the Flood Vulnerability Atlas.

2.3.1.2.3. Critical areas. At this stage, from the data and maps produced in the previous steps, the critical areas for the occurrence of floods (either due to high exposure to events or a lack of preparation for dealing with them) were identified. From this summary, the preventive actions below were recommended, preferably in the form of nonstructural measures. These measures required the adequacy of the telemetric network of ANA, the states, National Institute of Meteorology (INMET) and National Institute for Space Research (INPE), and required the drafting of a reference manual providing guidance

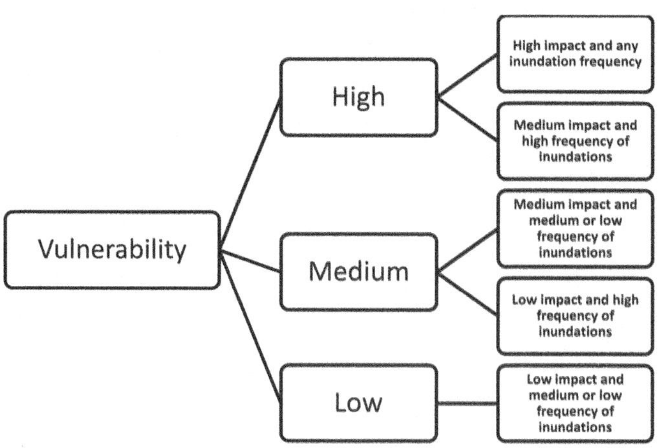

Fig. 6. Definition of the vulnerability level.

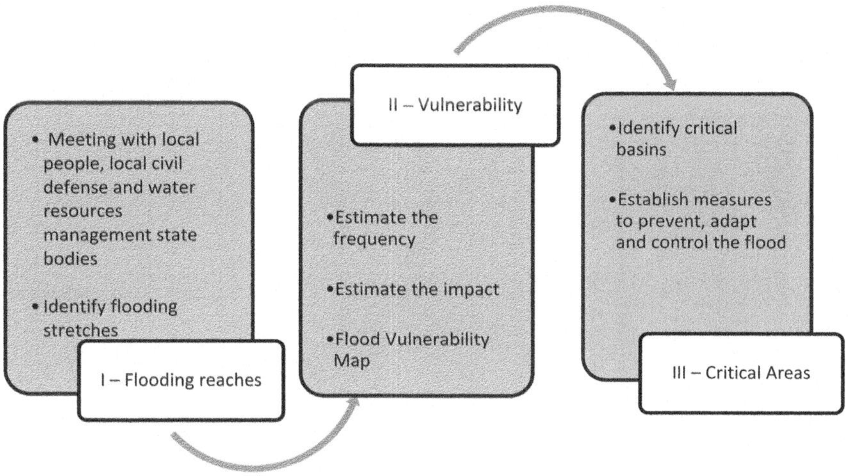

Fig. 7. Steps to build Flood Vulnerability Atlas.

on the technological and institutional options for coping with floods, involving nonstructural measures. According to the basin features, the following recommendations were made:

- Basins with elevated time of concentration:
  - stream flow forecasting models;
  - hydrologic warning systems.
- Basins with dense urban occupation:
  - charts of flood zones;
  - charts of flood risk.
- Basins with reduced time of concentration:
  - weather prediction systems.

As a final result, the Atlas presents maps showing the vulnerability of river stretches, such as the example shown in Figure 8.

The Flood Vulnerability Atlas identified 13,948 river stretches prone to flood in 2,780 rivers in the country, of which 4,111 stretches, or 30%, were considered highly vulnerable to gradual floods, 6,051 (43%) were of medium and 3,786 (27%) were of low vulnerability to such occurrences.

The regions that showed the highest proportion of stretches of rivers with high vulnerability to flooding are the midwest (45%) and the south (43%), while the others had less than 30% of their stretches with high vulnerability to gradual floods.

The north and northeast regions, although having significant gradual floods events, were the regions which were identified as having the fewest stretches being highly vulnerable to flooding. The north region has 82% of the sections identified with medium or low vulnerability to flooding and the northeast, 78%.

Owing to the large amount of maps obtained, it was necessary to simplify the presentation of the results, which was done in Table 1. This table shows, for each state, the main rivers with stretches of high vulnerability to flood. It is noteworthy that some critical sections which are smaller affluents are not listed, although several river stretches located in urban areas contribute to flood.

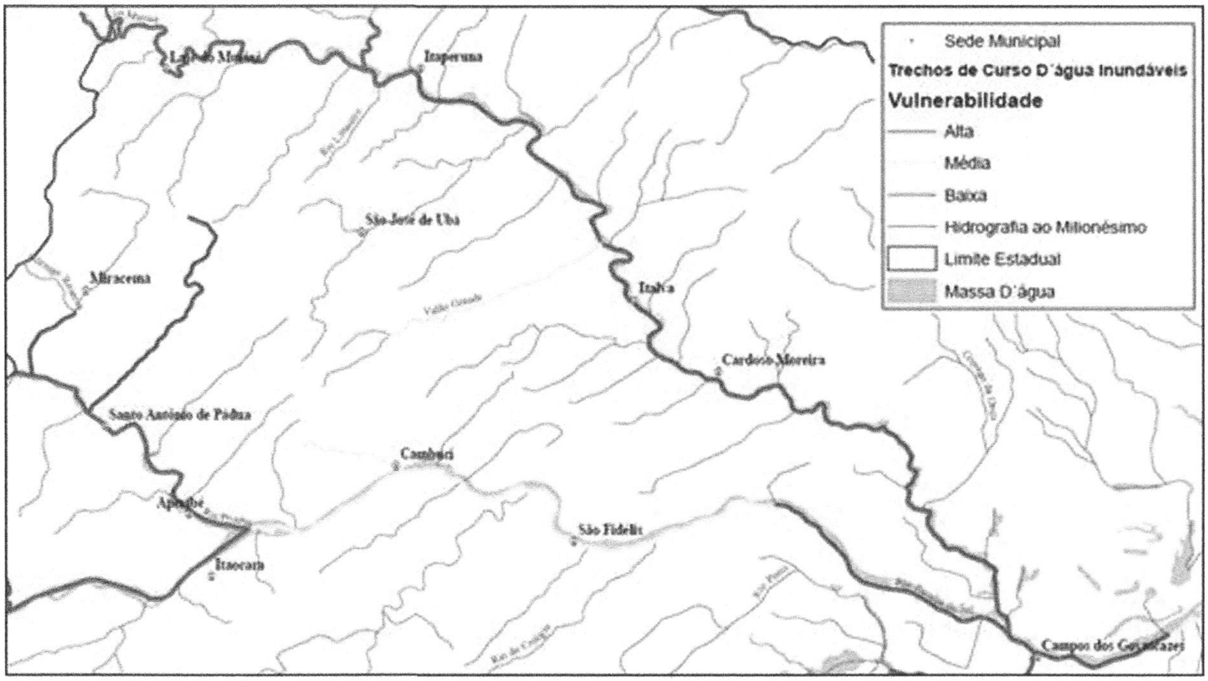

Fig. 8. Example of vulnerability map (Atlas of Vulnerability to Flooding).

As expected, the south and southeast regions are those with the largest number of rivers with vulnerability to flooding because their populations occupy a larger portion of the territory, settling in many cases in the lowland regions.

2.3.2. Priority watershed. The ANA Situation Room has some priority basins where there is frequent river level monitoring, especially in times of flooding, as in the case of the Doce, Mundaú and Acre rivers basins. The ANA Situation Room also monitors strategic basins for power generation, which is the case for the São Francisco, Large and Paraguay basins, or basins where important transpositions of flows occur, as in the cases of the Paraíba do Sul and Piracicaba rivers basins, among others.

As part of this monitoring, ANA issues daily or sporadic bulletins, depending on the hydrological situation set in the basin. The decision about the report period of a seasonal character bulletin is usually made based on the update of retention curves on existing stations in the basin, on the information available, and on the weather. To define new priority basins it is essential that Situation Room operators are guided by the results presented in the Flood Vulnerability Atlas.

2.4. Hydrometeorological stations

ANA is responsible for coordinating the activities carried out under the National Hydrometeorological Network, composed of more than 4,500 rainfall and river flow stations, where river level and river flow are monitored, as well as the amount of sediment and water quality. The rivers where monitoring

Table 1. Major rivers with stretches of high vulnerability to flooding.

Region	State	River
Midwest	Distrito Federal	Just small tributaries
	Goiás	Araguaia river, Ponte de Pedra river, Ribeirão Santa Maria river, Meia Ponte river, Paranã river and Vermelho river
	Mato Grosso	Cuiabá river and Araguaia river
	Mato Grosso do Sul	Paraguai river, Miranda river, Aquidauana river, Itiquira river, Apa river, Pardo river and Verde river
Northeast	Alagoas	Mundaú river, Paraíba river, Ipanema river, São Miguel river and Jacuípe river
	Bahia	São Francisco river, Itapicuru river, Jequirica river, Rio de Contas river, Cachoeira river, Almada river and Paraguaçu river
	Ceará	Acaraú river, Aracatiaçu river, Quixeramobim river, Cocó river and Jaguaribe river
	Maranhão	Itapicuru river, Mearim river, Munim river and Tocantins river
	Paraíba	Ingá river and Paraíba river
	Pernambuco	Una river, Pirangi river, Panelas river, Sirinhaém river, Jacuípe river, Mundaú river, Ipojuca river, Tapacurá river, Jaboatão river and Sirigi river
	Piauí	Parnaíba river and Gurguéia river
	Rio Grande do Norte	Apodi river, Piranhas-Açu river and Seridó river
	Sergipe	Caiçai river, Vaza-Barris river and Piautinga river
North	Acre	Juruá river, Tarauacá river, Envira river, Purus river, Iaco river and Acre river
	Amapá	Jari river, Araguari river and Amazonas river
	Amazonas	Solimões river, Negro river, Amazonas river, Madeira river, Purus river and Juruá river
	Pará	Amazonas river, Tapajós river, Xingu river and Tocantins river
	Rondônia	Madeira river, Mamoré river and Machado river
	Roraima	Branco river, Juaperi river, Igarapé Caracanã river and Igarapé Quitauaú river
	Tocantins	Araguaia river, Tocantins river, Manuel Alves Grande river, Lontra river, Ribeirão Tranqueira river and Formoso river
Southeast	Espírito Santo	São Domingos stream, Cricaré river, Cotaxé river, São Francisco river, São Mateus river, Doce river, São José river, Bananal river, Piraquê-Açu river, Fundão river, Guandu river, Santa Maria river, Jacu river, Santa Clara river, Castelo river, Itapemirim river, Muquiqui do Norte river, Muqui do Sul river, Rio do Veado river and Itabapoana river
	Minas Gerais	Guavinipã river, Paracatu, Mucuri river, São Nicolau river, Doce river, Piranga river, Xopotó river, Piracicaba river, Caratinga river, Matipó river, Mutum river, José Pedro river, São João river, Carangola river, Muriaé river, Turvo river, Maranhão river, Brumado river, Camapuã river, Paraopeba river, Jacaré river, Rio das Velhas river, Grande river and Rio das Mortes river
	Rio de Janeiro	Itabapoana river, Carangola river, river Muriaé, Pomba river, Paraíba do Sul river, Ururaí river, Canal da Andressa river, Macaé river, Preto river, Grande river, Preto river, Paquequer river, Capivari river, Tanguá river, Bananal river, Guandu river, Mambucaba river, Santana river and Sarapuí river
	São Paulo	Jacuí river, Jaguari river, Tietê river, Guanhanha river, Pariqueraçu river, Jacupiranga river and Ribeira do Igauapé river
South	Paraná	Tibaji river, Rio das Cinzas river, Pescaria river, Jaguaricatú river, Pitanga river, Nhundiaquara river, Atuba river, Belém river, Iguaçu river, Ribeirão dos Padilhas river, Barigui river, Rio da Várzea river, Negro river e Marrecas river

(*Continued.*)

Table 1. (*Continued.*)

Region	State	River
	Rio Grande do Sul	Uruguai river, Ijuizinho river, Ijuí river, Ligeiro river, Inhandava river, Carazinho river, Jacuí river, Soturno river, Vacacaí-Mirim river, Vacacaí river, São Sepé river, Arroio do Conde stream, Arroio dos Ratos stream, Pardo river, Zeferino stream, Taquari river, Santa Cruz stream, Caí river, Rolante river, Rio dos Sinos river, Gravataí river, Maquiné river, Camaquã river, Duro stream, Velhaco stream, Grande stream, Piratini river, Jaguarão river, Santa Maria river, Jaguari river, Ibicuí river, Ibiraputã river and Quaraí river
	Santa Catarina	Uruguai river, Rio das Almas river, Iraceminha river, Xanxerê river, Riacho Grande river, Santo Antônio river, Rio do Peixe river, Erval river, Canoas river, Guará river, Iguaçu river, Canoinhas river, Vermelho river, Pitanga river, Negrinho river, Cachoeira river, Texto river, Benedito river, Luís Alves river, Rio dos Índios river, Itajaí-açu river, Neise stream, Itajaí-mirim river, Taió river, Blumenau river, Itajaí do Oeste river, Itajaí do Sul river, Alto Braço river, Rio das Antas river, Cubatão river, Biguaçu river, Rio do Meio river, Braço do Norte river, Tubarão river, Mãe Luzia river, Manoel Alves river, Araranguá river, Sertão river and Mampituba river

occurs correspond to 2,176 of the 12,978 registered rivers in ANA's hydrological information system ANA[2].

This information is critical for both ANA's water resources management decision making as well as the development of projects in various sectors of the economy that are water users, such as agriculture, water transportation, hydroelectric power generation, sanitation and aquaculture.

In recent years, ANA has invested in the modernization of the hydrometeorological network, installing telemetric stations which, through the DCPs, carry out the automated acquisition of hydrologic data and transmit them to the Agency, where the data are processed, stored and made available online.

This type of equipment has several advantages, such as (i) it allows monitoring in areas of difficult access, (ii) it enables the monitoring in real time of critical hydrological events as well as the volume stored in reservoirs, (iii) it feeds water quality warning systems, and so on. For this reason, ANA began to adopt the telemetric stations as a reference in planning the expansion of the hydrometeorological network under its responsibility.

2.5. *Reservoirs*

Reservoirs are bodies of water, natural or artificial, used to store, regulate and control water resources. The basic purpose of storage is to retain excess water from the wet season to ensure water reserves in the dry season. Meanwhile, regularization corresponds to the quantity of water that a reservoir can provide in a given period of time.

The control of water resources is the operational aspect that concerns the way that water is used, as it can be stored or released, and how that should be achieved. The control is achieved by the operation of

[2] ANA provides data level, flow, sediment and water quality in Brazilian rivers and rainfall in the country at the following sites: HIDROWEB http://hidroweb.ana.gov.br/; hydrological monitoring system http://www.ana.gov.br/telemetria; and National Information System on Water Resources – SNIRH http://www.snirh.gov.br/.

the reservoir which relies on the definition of operating rules regarding the level of water that the dam must keep and flow rates to be flushed downstream. The level is directly associated with the stored water volume, which can be used for multiple purposes: human consumption, animal supply, irrigation, power generation, aquaculture, industrial use, flood control, etc. The flow released downstream may also be related to uses that may be downstream, including environmental water use to preserve the organisms that depend on it.

Regarding critical hydrological events, a high water level can cause backwater upstream, i.e., super-elevation of the water level of the river, flooding upstream regions. A low water level, in turn, reduces the ability to stabilize the reservoir, which can characterize a period of water shortage. Moreover, in the rainy season, it is possible to reserve part of the volume of the reservoir to retain a predicted flood wave.

In these critical situations of floods and shortage of water, the reservoir also has significant relevance to downstream areas. Flows released can soften the impact of floods, in that they reduce the natural flow spillover to the limit of the river channel, or relieve the pressures on water resources. On the other hand, it can increase the water supply by releasing higher flow during the drought season.

In this context, ANA has an important role, since it has, as one of its tasks, to define and monitor the conditions of operation of reservoirs by public and private actors, to ensure multiple use of water resources (as established in the resource water plans in their basins). In hydropower exploitations, ANA negotiates the definition of reservoir operating conditions with ONS.

ONS annually consolidates and offers an 'inventory of operative constraints of hydroelectric drain' on its homepage, providing information about hydraulic operational constraints originating from surveys conducted in the past and periodically, relative to the minimum and maximum flows sections and river stretches, limitations of maximum and minimum flows in exploitations, maximum and minimum reservoir levels, maximum rates of change of hydraulic flows, and other restrictions. This inventory also shows a schematic diagram of the power plants of the National interconnected system (SIN), classifying by basin.

ONS also annually prepares and provides an 'annual flood prevention plan', which contains the results of the studies conducted to define the volumes expected to be maintained in the associated reservoirs with different hydrological scenarios grouped by basin. The flood-control volume corresponds to the portion of the net volume of the dam to be maintained during the period when aiming to retain part of its volume, protecting downstream areas because the river flow can be reduced.

2.5.1. Definition of reservoirs for critical events monitoring. The definition of the reservoirs must take into account hydrological peculiarities of the region and the relative importance that they have: in times of shortage, regulating reservoirs are strategic to meet water demands; in the wet seasons, reservoirs with a flood-control volume and the damping capacity of the flow should be considered in controlling floods.

Run-of-the-rivers reservoirs are those which make little alteration of natural river flows, with less relevant flood control. However, knowledge of the characteristics of the operation and monitoring of these are needed, since these are constructions that interfere with the natural flow.

Table 2 lists the main features to be observed in defining reservoirs to monitor the activity of critical hydrological events, monitoring scarcity of water and flooding. Moreover, it presents some important issues to be monitored if there is available information.

In general, the larger reservoirs in a basin are used both in ensuring water supply in times of scarcity and in flood control. In Brazil, only the reservoirs linked to the SIN hold most of the information in a

Table 2. Definition of reservoirs for critical events monitoring.

Period	Main characteristic	Relevant information
Dry	Storage capacity; Regularization capacity;	Storage volume; Target volume of the period; Minimum flow released downstream; Maximum water withdrawals for the period; Climate forecast.
Wet	Total hold volume; Cushioning capacity of flood flows;	Reservoir level; Target level of flood control storage volume; Expected inflow; Predicted outflow; Maximum outflow; Weather forecast.

systematic way. This means that in most reservoirs for multipurpose uses that are not required for the generation of hydropower, information has to be raised from various sources or generated from specific studies.

The National Electric Energy Agency provides the georeferenced information system for the electricity sector where data about the exploitations of the power sector can be obtained. In addition, some other data can be obtained from the websites of State Water Resources Managers Agencies.

2.5.2. Characterization of reservoir operation situations. The characterization of reservoir operation for flood control should consider the occupation of flood control storage, the forecasted inflow and outflow and maximum outflow, which is usually associated with the flow limit supported by the riverbed at critical points downstream.

The main differences in the characterization of scarcity in relation to flooding is that the length of the first is longer, typically in the order of months, and its onset occurs when water scarcity compromises the burden of water demands, especially of water supply systems. Thus, scarcity is characterized from the average inflow, the reservoir level (which is associated with a storage volume) and the planned rate of withdrawal flow.

It is noteworthy that the operating rules may be disregarded in emergency situations, when there is characterized imminent risk to the health of the population, the environment and hydraulic structures due to accidents or floods. In these cases, it is recommended that the system operations are carried out by the operator, with the accompaniment of the bodies involved – management agency, committee and so on.

3. Situation Room Action Plan

An annual Situation Room Action Plan should be prepared to guide its operation, indicating minimally:

- regions or priority basins to be monitored during the period of the Plan, considering the critical event;
- actions in the Situation Room to be developed by region or basin and its stage in the development of each action;

- staff available and the allocation of activities among its members, considering the technological resources available.

In general, the Situation Room Action Plan translates into the generation and dissemination of information on critical hydrological events.

4. Final considerations

Since the implementation of the National Civil Defense System in Brazil in 1988, the focus on disaster management is no longer the response when assisting those affected; it has now become managing disaster risk, in addition to the response, including prevention and minimizing the effects of the critical event.

ANA, established in 2000, has been engaging in risk management, with the pillars of hydrological monitoring, mapping risk areas and improved responsiveness to the occurrence of extreme flood and drought events.

In this context, we highlight the actions for modernization and expansion of the hydrological monitoring network, in particular the alert network, the completion of the Flood Vulnerability Atlas and deployment of Situation Rooms in the states, with a view to strengthening the role of these bodies in disaster risk management and providing them with greater capacity to respond to such events.

In recent years, a growing concern regarding risk identification and prevention of natural disasters has been observed in Brazil. In 2013, 273 emergency decrees (SE) were published or states of emergency (ECP) declared due to the occurrence of floods in 262 municipalities (for approximately 4.7% of all municipalities in the country, the issuing of decrees related to these types of events has been decreasing since 2009 and, in 2013, the number of decrees issued was registered as the fourth smallest value in a series of eleven). Although it may mean that the events are more localized, this information also reflects the effectiveness of the actions that have been implemented by the bodies in the National System of Civil Protection and Defense, among them the ANA. Certainly, the impacts of extreme weather events cannot be eliminated but, through prevention, forecasting and warning, it is possible to reduce the damage to property and infrastructure, and the loss of human lives.

References

Brazil Ministry of National Integration & Federal University of Santa Catarina (2012). *Brazilian Atlas of Natural Disaster: 1991–2010*. Volume: Brazil. Ministry of National Integration & Universidade Federal de Santa Catarina (UFSC), Florianópolis.

Brazil Ministry of Planning (2012). National Risk Management and Disaster Response Plan: 2012, 2014 - Presentation. http://www.planejamento.gov.br/secretarias/upload/Arquivos/pronunciamentos/apresentacaoes/2012/120808_Plano_Nac_Risco_2.pdf.

Water-related disaster risk reduction: time for preventive action! Position paper of the High Level Experts and Leaders Panel (HELP) on water and disasters

Koos Wieriks[a,b,*] and Niels Vlaanderen[a]

[a]*Ministry of Infrastructure and the Environment, Den Haag, The Netherlands*

[b]*UN Secretary General's Advisory Board on Water and Sanitation*

**Corresponding author. E-mail: koos.wieriks@minienm.nl*

Abstract

Climate change is exacerbating the extremes in hydro-meteorological events. Together with other global drivers under change – population growth, rapid urbanisation, increased asset values – this may result in increased frequencies and even higher impacts of water-related disasters. Further comprehensive actions are needed to reduce the vulnerability and to increase the resilience of exposed populations and assets.

Keywords: Delta approach; Disaster risk reduction; Floods; High Level Expert and Leaders Panel (HELP); Resilience; The Netherlands; Water-related disaster

Messages

1. Water-related disasters, such as floods, droughts, storm surges and tsunamis account for 90% of all disasters in terms of number of people affected. This number is increasing. The poor, vulnerable groups, women and girls are suffering the most. Economic and environmental losses associated with water-related hazards are rising in all regions.
2. Disaster risk reduction, water resources management and climate adaptation should no longer be treated as separate topics.
3. More data and better tools for risk assessment are ready for use now and need to be more widely deployed to identify and prioritise actions. Better preparedness of citizens in terms of risk awareness and emergency planning are also essential.
4. Risk reduction, preparation and prevention are sensible investments that pay off in terms of reduced loss of life, avoided damage, and long-term economic growth and stability. Further emphasis on the role of financial protection measures in disaster risk management (DRM) is needed.

doi: 10.2166/wp.2015.011

5. Risk prevention should be integrated with long-term planning. This allows communities and decision makers to identify and exploit opportunities for synergies with planned investments, including plans for adaptation to climate change.
6. Uncertainties are no excuse for inaction: uncertainties are inherent in long-term planning and should be accounted for in a comprehensive, flexible and adaptive approach.
7. The efforts under each of the major 2015 international policy frameworks to create synergy and to increase effectiveness should be aligned.

1. Climate shifts and other global changes are already impacting water-related disasters

The projected impacts of climate change clearly indicate consequences for the occurrence of disasters: enhanced sea level rise and more pronounced hydro-meteorological extremes, with a higher frequency of intense storms, locally more intense rainfall, higher river discharge extremes, but also longer dry periods and droughts that can lead to deterioration of already scarce water resources.

In some regions, new kinds of disasters are occurring, such as droughts, in areas that have not experienced such kinds of impacts before.

Meanwhile, other global drivers contribute to the increase in the vulnerability of the population, resulting in potential additional risks and to the casualties and hazards that may occur in the case of extreme events. Population growth, asset deterioration, rapid urbanisation and subsidence from groundwater extraction may pose an additional risk to already vulnerable urbanised flood-prone areas, as does the increased value of assets which are often concentrated in and around these same urban areas.

1.1. Water-related disasters have large impacts and these impacts are growing in scope and severity

Globally, water-related disasters already account for 90% of all natural disasters[1]. Their frequency and intensity is generally rising due to climate change, causing enormous damage to life and property. Climate change is a factor in these trends. Damages attributed to water-related disasters can amount to 15% of annual gross domestic product (GDP) for certain countries.

Population growth, poverty, land shortages, urbanisation, the poor condition of flood protection and drainage infrastructure, and water storage facilities, especially in developing countries, have increased the vulnerability of people to flood hazards and droughts, and, *inter alia*, have multiplied impacts on public health associated with waterborne epidemics. The poor, vulnerable groups, women and girls are suffering the most.

Moreover, droughts, as slowly developing disasters, but large in extent, lead to the collapse of social structures and to refugees that may cause disruptions in social structures of adjacent regions.

Recent history has taught us that even the most developed countries are vulnerable to water-related hazards. Climate change is very likely to exacerbate this trend, both for floods and droughts, although in different parts of the world different trends in precipitation and temperatures are expected.

---

[1] United Nations Educational, Scientific and Cultural Organization (UNESCO) (2012). *Managing Water under Risk and Uncertainty: The United Nations World Water Development Report 4*. UNESCO, Paris. http://unesdoc.unesco.org/images/0021/002156/215644e.pdf.

2. Disaster risk reduction, water resources management and climate adaptation should no longer be treated as separate topics

To date, most disaster response agencies at global, national and local levels have treated disaster threats as (a) relatively fixed and unchanging and (b) threats that should be responded to after their occurrence rather than in advance through planning and preparation.

Indeed, in terms of development aid, emergency aid, global adaptation funding, and national water and emergency planning processes, the divisions between climate adaptation, water resources management and disaster response seem potentially dangerous, exacerbating risks to communities, economies and the environment.

Given the shifting nature of water-related threats, climate adaptation, water resources management and disaster risk reduction should be merged and better integrated to address new and emerging impacts.

3. Information to manage and predict water-related disasters needs improvement

Data on the impacts of water-related disasters (floods and droughts) are increasingly available but vary considerably in quality and quantity. Assessing vulnerability is an essential tool in water-related disaster management. Yet, documentation of disease outbreaks and public health after-effects of water-related hazards is still lacking.

Another limitation is our understanding of how to account for secondary (or indirect) economic consequences, such as long-term disruptions to economic chains, economic damage through infrastructure impact, and environmental damage from increasing competition between increasing water resources demands from human communities and activities and ecosystems.

Greater consistency in the reporting and documenting of water-related disasters, as well as a better understanding of impacts based on common criteria, are crucial to establish baselines, set priorities, track trends and assess the effectiveness (costs and benefits) of any proposed response.

4. Disaster management as currently practised is typically reactive, rather than proactive

Historically, water-related disasters have been thought of and treated as isolated, one-off events that triggered responses in their aftermath as reactive crisis management (relief and response). Hence, disaster risk reduction has often focused on improving primary, reactive responses, such as capacity building for civil defence on how to cope with major disasters such as tsunamis, or developing drought or flood contingency plans on how to coordinate emergency responses among a variety of actors. Compared to relief and response, much less attention has been paid to prevention and mitigation.

4.1. Shifting disaster management from reaction to prevention

Preventive action aims at developing measures to prevent disasters from happening or measures to increase the resilience (see Box 1) to cope with potentially disastrous events.

Box 1. Resilience, vulnerability and risk assessment

Resilience can be described as 'the ability of a system, community or society exposed to hazards to resist, absorb, accommodate to and recover from the effects of a hazard in a timely and efficient manner, including through the preservation and restoration of its essential basic structures and functions' (UNISDR Terminology on Disaster Risk Reduction (2009); see: http://www.unisdr.org/files/7817_UNISDRTerminologyEnglish.pdf).

Climate adaptation assumes that, under novel conditions, rather than just recovery, the anticipation of desired new socio-economic developments is also an inherent element of resilience.

Resilience concepts need to be further developed for critical infrastructures (supply of basic services like water, food, energy, transport, housing/shelter, communications, finance, health), but also for the wider public, to integrate and address human and social dynamics in crises and disaster situations.

Resilience concepts take into account the necessity to anticipate, to plan and to implement in the crises time a substitution process aiming to deal with a lack of material, technical or human resources or capacities necessary to assume the continuity of basic functions and services until recovery from negative effects and until return to the nominal position. Immediate civil organisational arrangements post-disaster are essential.

Moreover, as resilience management and vulnerability reduction are closely related, it is necessary to link the on-going efforts and share risk assessment and mapping approaches, e.g. physical exposure mapping with relevant resilience management approaches, to ensure that risk assessment is followed by the development of resilience concepts in the various security sectors, based on the results of the risk assessments.

Several examples exist to alleviate droughts by, for instance, the use of aquifer storage and recovery or building sand dams, or by better forecasting and early warning (see Box 2), or to address flood issues through integrating 'green infrastructure' in management plans (Box 3) and by addressing climate uncertainties through more flexible management systems (Box 4).

This approach pays: risk prevention is usually a sensible investment. For example, some studies show that in certain situations every euro/dollar spent on preventive measures can pay back up to ten-fold in avoided damage and loss of life. As the incidence and severity of extreme events are expected to continue to increase, investment in prevention is becoming increasingly advantageous. These observations are not new. In 2005, the Hyogo Framework for Action articulated the need for a more risk based and preventive approach, though little progress has been observed in this area in response to this agreement. This has to be considered against an underlying problem of deterioration of existing assets and asset systems due to long-term underinvestment around the world.

4.2. *Incorporating financial protection as a measure and an incentive for preparedness*

Financial protection can serve as an effective preparedness measure, which can drastically reduce the impact of disasters. For instance, the World Bank has undertaken a pilot index-based livestock insurance programme in four counties in Northern Kenya, which builds on Kenya's Hunger Social Safety Net Programme, with the aim of scaling up to a large-scale livestock insurance programme. The primary

Box 2. Strengthen preparedness to droughts

By advancing drought forecasting, early warning and mitigation practices, the drought early warning and forecasting in Africa (DEWFORA) project aims to reduce vulnerability and strengthen preparedness to droughts in Africa. To this aim, a framework for monitoring, predicting, timely warning and responding to droughts at the seasonal time scale is being developed, applicable within the institutional context of African countries. The project also contributes to improve the identification of vulnerable regions taking into account the increased hazard due to climate change, and has developed feasible adaptation measures.

The project shows that improved drought forecasting and early warning based in vulnerability assessments can improve the preparedness to droughts in Africa. However, lack of financial resources, and institutional arrangements that are not suited for the operation of an early warning system and dissemination of its results, lead to a situation in which the available information is not suitable and not timely for the end-users and is generally unreliable and inadequate.

In collaboration with the United Nations Educational, Scientific and Cultural Organization's International Hydrological Programme, Princeton University has developed an experimental drought monitoring and forecast system for sub-Saharan Africa, since current approaches to drought monitoring in developing regions have generally been limited, in part because of unreliable monitoring networks and limited national capacity. Hydrological and drought forecasts are provided out to 6 months (http://hydrology.princeton.edu/monitor). The Africa Drought Monitor (ADM) has been implemented at the AGRHYMET (Centre Regional de Formation et d'Application en Agrométéorologie et Hydrologie Opérationnelle) regional centre in Niamey, Niger, and at the Greater Horn of Africa Region at the Intergovernmental Authority on Development Climate Prediction and Applications Centre in Nairobi.

Source: Dewfora (http://www.dewfora.net/index.php).

objective of the programme is to reduce the vulnerability of pastoralist households against drought, an effective risk retention and preparedness measure. Furthermore, the bank also provides contingent instruments (such as World Bank Catastrophe Deferred Drawdown Options) which are pre-approved credit lines that offer immediate sources of liquidity to countries when disasters strike. For countries to be eligible for these risk transfer mechanisms, they must implement a comprehensive DRM programme, which the bank monitors on a periodic basis. Therefore, these credit lines serve as incentives to incorporate effective preparedness measures in national level strategies. In addition, the DRM programme helps countries identify and understand risk which can then prompt other effective prevention measures.

5. How do we include preventive action in disaster risk reduction?

In recent years an integrated, adaptive approach has been developed based on the recognition that risks, if not properly addressed in advance, may lead to the occurrence of disasters. Its widespread and effective implementation requires a reframing of the paradigm of DRM from its traditional focus on mitigating the (direct) impacts of disasters using stand-alone and *ad hoc* interventions to encompass:

Box 3. Green infrastructure for disaster risk reduction in the Mekong Delta, Vietnam: merging traditional and innovative solutions

Floods play an important role in the life of people living in the Mekong Delta. Each year floodwaters inundate 1.9 million ha (19,000 km^2) and affect the lives of more than 2 million people. Normally, these floods are essential to food security and biodiversity and people have a tradition of living with the floods. However, extreme flood events can be destructive and cause enormous damage.

Sea level rise is expected to result in large areas of more frequently and even permanently inundated coastal plains. Furthermore, sea level rise will increase salinity levels in the delta rivers and its water network. Agricultural production will be affected through more frequent and longer periods of flooding, and as a result of salinity intrusion. The population size is projected to almost double by the year 2050 from 17 million to around 30 million. This will fuel the urbanisation trend, taking more land out of agricultural production. At the same time more people need to be provided with food and fresh water. Ongoing industrialisation will also take up more space and increase the demand for water as well as the production of wastewater. Both trends will increase the need for proper spatial planning, efficient water supply, investments in water treatment and stringent enforcement of environmental legislation.

The current water management system is based on the 1994 Mekong Delta Master Plan. An update is presently under discussion in order to include environmental and socio-economic developments as well as recent innovations and modern approaches, such as green water defences. This concept can be very useful in combination with the traditional measures for flood control. In fact, several good practices of the concept can be found in Government decisions and design guidelines for infrastructure and land use.

Source: Li, X., Marchand, M. & Li, W. (2012). *Grow in Concert with Nature: Green Water Defence for Flood Risk Management in East Asia. An EASSD Discussion Paper.* World Bank, Washington DC.

(i) a broader focus on prevention, mitigation, preparedness and vulnerability reduction; (ii) the recognition that DRM is surrounded by many uncertainties and should therefore be an iterative or continuous, on-going effort requiring experimentation and learning; and (iii) the integration and mainstreaming into sustainable development policies, planning and programming at all levels, globally and nationally as well as locally and at community level.

As an example, spatial planning should include considering restrictions for building in areas exposed to disaster impacts, thus reducing exposure of people and assets. UNISDR currently facilitates the development of a post-2015 framework for disaster risk reduction. This offers an excellent opportunity to include these latest experiences and insights in disaster risk reduction. The post-2015 framework should therefore have a stronger foundation through a risk based, preventive approach. Such an approach should be accompanied by agreements on concrete actions as well as monitoring of such actions.

5.1. Preventive action is an opportunity for (economic) development

There are many opportunities for reducing water-related disaster vulnerabilities in the face of global change. While aging infrastructure and building stock in the developed world pose a risk due to increasing vulnerability, this also provides an opportunity to introduce new technologies in the redevelopment process and to adapt infrastructure and buildings to enhance disaster resilience.

Box 4. Adaptive delta management: optimising investments in risk reduction

In the Netherlands, the central government, water boards, provinces and municipalities are working together on a new Delta Plan on Water Risk Management. This programme is referred to as the Delta Programme. Its primary goal is to protect the Netherlands against floods and ensure the availability of fresh water, now and for future generations.

The Delta Programme comprises a cohesive set of projects (measures) for the short term, but also looks ahead to the medium and long term (up to 2050). The programme has developed a new, adaptive management strategy: the adaptive delta management (ADM) approach. ADM is defined as 'a smart and intelligent way of taking account of uncertainties and dependencies in decision-making on delta management with a view to reducing the risk of overspending or underinvestment'. ADM entails a phased approach towards investments while decision-making is driven by major uncertainties around future developments and the desirability of responsible financial investment. It starts off from short-term decisions in the broad fields of water, land use and spatial planning and links to long-term issues in the specific fields of protection against flooding and freshwater supply, allowing switching between strategies through adaptation pathways. Short-term measures must be logical in the long term: they are useful, do not obstruct long-term measures, or are even necessary to keep long-term options open.

Important features of the Delta Programme are:

(1) involving multiple stakeholders in a joint decision-making process to enhance legitimacy and feasibility;

(2) taking a risk-based perspective;

(3) adopting a flexible approach in the possible strategies by valuing flexibility with regard to the timing of implementation;

(4) inter-linking various investment agendas and looking for opportunities for mainstreaming with planned investments; and

(5) other key success factors, including the Delta Programme being led by an independent coordinator; considering the financing of both planning and implementation; and having a firm legal basis.

Source: Dutch Delta Commissioner (http://www.deltacommissaris.nl/english/).

Urban restoration, regeneration and modernisation can be a key driver of economic development, both as a result of the initial investments required and the benefits that will accrue over time (e.g. formerly flood-prone areas may become available for productive use).

In addition, in many new economies and developing countries, their growing economy also provides an opportunity to use lessons from the developed world to avoid some past mistakes (leap frogging).

Better preparedness of the civil population through awareness-raising and clear actions to be taken in advance of or during a disaster can also significantly reduce the human impact.

6. Increased resilience is needed to deal with uncertainties

Uncertainties are no excuse for inaction; they are inherent in long-term planning and should be accounted for in an adaptive approach, which may be adjusted over time in response to evolving economic and environmental conditions.

There is an urgent need to enhance the disaster resilience (both conditions and performance, see also Box 1) of our critical infrastructure using preventive measures, so as to secure the basic supply of services such as water, food, energy, transport, housing, communications, finance and health. This should take into account the various interactions between sectors at the local scale and also how these interact at regional, national or even transboundary levels.

6.1. Preventive action is not to be executed in isolation

Clearly, preventive action against disasters requires a long-term development perspective. It can therefore not be linked to DRM alone. Preventive action needs to be embedded in an integrated process that incorporates water resources management and climate adaptation processes (see also Section 2), and mainstreamed into national planning processes.

Such an approach also enables optimisation of resources, where, for instance, flood prevention infrastructure can also serve the purpose of drought prevention, and at the same time hydropower and/or irrigation.

6.2. Towards a flexible approach

There is growing international recognition that water-related DRM should be a programmed and flexible process of continuously improving management practices, in which short-term actions are linked to long-term goals, flexibility is valued and incorporated, multiple strategies are considered in a rational manner and different investment agendas are inter-linked.

The development of a post-2015 framework for disaster risk reduction offers an excellent opportunity to put this recognition into practice, developing a framework that builds on preventive action and promotes flexible implementation of measures.

7. Align efforts under the major 2015 international policy frameworks to create synergy and increase effectiveness

In 2015, a number of important international policy framework events take place, including the decision-making for the post 2015 Sustainable Development Goals, the Sendai World Conference on DRM (the Hyogo Framework for Action), and the Paris Climate Conference. By aligning these frameworks in terms of DRM, synergy and efficiency can be achieved, for instance in tracking and measuring goals and outcomes.

IWA Publishing's authorised EU representative for General Product Safety Regulations is Diane D'Arras, 15 rue Duret, 75116 Paris, France, e-mail: safety@iwap.co.uk.

Printed and bound by CPI Group (UK) Ltd, Croydon, CR0 4YY
12/05/2026
02108880-0019